UNIVERSITY PRESS OF FLORIDA

Florida A&M University, Tallahassee
Florida Atlantic University, Boca Raton
Florida Gulf Coast University, Ft. Myers
Florida International University, Miami
Florida State University, Tallahassee
New College of Florida, Sarasota
University of Central Florida, Orlando
University of Florida, Gainesville
University of North Florida, Jacksonville
University of South Florida, Tampa
University of West Florida, Pensacola

WILLY LEY

Prophet of the Space Age

Jared S. Buss

University Press of Florida
Gainesville · Tallahassee · Tampa · Boca Raton
Pensacola · Orlando · Miami · Jacksonville · Ft. Myers · Sarasota

First cloth printing, 2017
First paperback printing, 2020

25 24 23 22 21 20   6 5 4 3 2 1

Library of Congress Cataloging-in-Publication Data
    Names: Buss, Jared S., author.
    Title: Willy Ley : Prophet of the Space Age / Jared S. Buss.
    Description: Gainesville : University Press of Florida, 2017. | Includes
    bibliographical references and index.
    Identifiers: LCCN 2017005527 | ISBN 9780813054438 (cloth)
    ISBN 9780813068244 (pbk.)
    Subjects: LCSH: Ley, Willy, 1906–1969—Biography. | Rocketry—United
    States—Biography. | Scientists—United States—Biography.
    Classification: LCC TL781.85.L49 B87 2017 | DDC 629.4092 [B] —dc23
    LC record available at https://lccn.loc.gov/2017005527

The University Press of Florida is the scholarly publishing agency for the State
University System of Florida, comprising Florida A&M University, Florida Atlantic
University, Florida Gulf Coast University, Florida International University, Florida
State University, New College of Florida, University of Central Florida, University
of Florida, University of North Florida, University of South Florida, and University
of West Florida.

University Press of Florida
2046 NE Waldo Road
Suite 2100
Gainesville, FL 32609
http://upress.ufl.edu

apogee supported by a grant
Figure Foundation

to explain the near impossible

Contents

Figures

Acknowledgments

Many people have contributed advice and guidance during the research and writing of this book. I would like to thank especially colleagues at the Smithsonian National Air and Space Museum in Washington, D.C. Curator and historian Michael Neufeld supervised book revisions during a 2015 NASM Guggenheim fellowship. His insights and perspectives shaped this book in significant ways, and I am forever grateful for his time, patience, and encouragement. Other scholars at NASM provided feedback, advice, and corrections. In no particular order, I wish to thank Roger Launius, Martin Collins, David DeVorkin, and Tom Crouch. Each of these individuals provided feedback and support that became invaluable at different stages. I would like to offer special thanks to Margaret Weitekamp, who provided both valuable professional advice and personal assistance with the hellish daily commute in Washington, D.C. I greatly enjoyed spending time with her lovely family. Additionally, museum project specialist Hunter Hollins served as an extremely helpful reader and colleague. This book benefited from other helpful NASM library and archival staff, both downtown and at the Steven F. Udvar-Hazy Center.

Many colleagues guided the project during the research and early writing phases. Hunter Heyck provided his critical insights on biographical writing and Cold War science. Katherine Pandora offered invaluable expertise on American popular culture and popular science. JoAnn Palmeri shaped my perspectives on the history of astronomy, while Stephen Weldon provided critical insights on secular humanism during the twentieth century. I thank Benjamin Alpers for his particular insights on political discourse, popular culture, and the Cold War. As is often written, any mistakes in this book are my own mistakes. The staff of the University Press of Florida are wonderful.

I would also like to thank the History of Science Society and NASA for their early support through the HSS/NASA Fellowship in the History of Space Science. This fellowship greatly assisted with archival research in the Willy Ley Collection at Udvar-Hazy, the Pendray Papers at Princeton, and Robert A. Heinlein Archives. I owe a special thanks to Heinlein's biographer, the late William H. Patterson Jr., who greatly assisted me with the Heinlein Archives. Other thanks are due to archivist Anne Coleman, science fiction editor Wolfgang Both, and Bonestell LLC.

This book benefited from short and long conversations with Ley's daughters, Sandra Ley and Xenia Ley Parker. While Sandra provided valuable time for telephone interviews, Xenia greatly assisted with personal photographs, e-mail interviews, and other requests. I sincerely hope that this book honors and respects the memory and legacy of their father. Willy Ley became my historical subject, and I can never hope to know him as they did.

Lastly, I would like to thank my wife, Loraine, who has provided intellectual and emotional support throughout so many years of research and writing. She has been an amazing partner and friend. This book would not have been possible without her guidance and endless patience.

Abbreviations

ASF	*Astounding Science Fiction*
CDT	*Chicago Daily Tribune*
CS-T	*Chicago Sun-Times*
FPP	Frederik Pohl Papers, Syracuse University
HA	Robert A. and Virginia Heinlein Archives, University of California Santa Cruz
LAT	*Los Angeles Times*
NYT	*New York Times*
PP	G. Edward Pendray Papers, Mudd Manuscript Library, Princeton University
WP	*Washington Post*
WLC	Willy Ley Collection, Smithsonian National Air and Space Museum, Udvar-Hazy Center

Introduction

This book explores the life and works of Willy Ley (1906–1969), a German-born American writer who defies classification. Journalists called him a scientist, yet he rarely practiced science in an institutional or laboratory setting. Reviewers called him a rocket engineer, yet he rarely designed rockets. Science fiction writers labeled him a prophet who predicted the future, yet Ley often wrote as a historian of science looking to the past. Historians may call him a modernist who celebrated the conquest of nature, grand engineering redesigns, and the future. Simultaneously, Ley could be described as a romantic who searched for wholeness, experienced awe and wonder, and voiced nostalgia for a time when science was open to all.

It is most accurate to describe Willy Ley as a science writer and popularizer. He could also be called a science educator, so long as the realms of education are recognized as broad, popular, and media-savvy. He wrote or contributed to over fifty books, and his newspaper and magazine articles are too numerous to count. He also gave hundreds of public talks and interviews. He spent much of his life on lecture tours. Ley could be called a publicist, especially during the 1940s and early 1950s, when he became America's most prominent rocket expert. These activities as a spaceflight advocate have garnered Ley a prominent position in the history of spaceflight, due to his efforts jointly popularizing rockets and a future of interplanetary travel. He has also been credited as the first historian of spaceflight, inaugurating a field of scholarship that flourishes today.

If there is one label that Willy Ley embraced, it was that of a romantic naturalist. One might also use the label of modern romantic to describe him. As a naturalist who embraced wonder, awe, and the mysteries of nature, he simultaneously sought to unmask, conquer, and master nature. He used gendered terms to express his thoughts. Ley stood enchanted by the beauty and wonder of nature, just as he sought to possess and objectify "her" treasures. As a product of the early twentieth century, he recognized the continuity with the themes and goals of nineteenth-century explorers and popularizers. Often, he attempted to revive or reinvigorate a genre of popular science that combined science and imagination with a strong dose of art, speculation, and optimism about the future.

Like other popularizers, especially Carl Sagan, he was a man of contradictions. For example, he trusted eyewitness accounts of the abominable snowman, while he dismissed eyewitness accounts of UFOs. He celebrated the great unknowns of nature, while he relentlessly debunked the myths and legends surrounding great unknowns. He possessed an unflinching imagination, restrained by a skeptical mind. He celebrated great leaps into the unknowns of science, yet warred against the pitfalls of such intuitive leaps. He democratized science for all, while he cherished the utility of expertise. He embraced mass media, while he campaigned against bestselling books. His causes and rationales were inconsistent.

Ley's inconsistencies and contradictions are telling. His perspectives illustrate the complexities of a "modern form of enchantment." As scholar Michael Saler argued, we can recognize many writers and cultural producers who did not simply rebel against modernity in the early twentieth century. Instead, we can appreciate how certain writers sought to complement or alter the discourse. They sought to retain space for marvels, wonders, and even miracles. Yet they remained committed to secularism, rationality, and empiricism. They tapped into the widespread need for spirituality that meshed with empiricism, reason, and science. Their works offered transcendental meaning and reverence for the great unknowns. They defended intuition, imagination, and even a spiritual conception of the cosmos. They approached scientific mysteries and new frontiers with open minds.[1]

Modern science (or Western thought) has often been presented as the central force of disenchantment, as scientific thinkers waged wars against superstition, mystical thinking, revelation, and magic. According to traditional narratives, an aggressive, mechanistic, or secular worldview stripped away the mysteries to reveal a universe destined for conquest and control. Reason and empiricism triumphed over intuition and revelation. In the famous perspective of sociologist Max Weber, modern life offered an iron cage, barren and soulless. Wonder and surprise became the relics of a medieval past. These anachronisms could also be seen as living fossils, now gasping for air in vulgar, profane realms of popular culture. In traditional accounts, magic would be replaced by rationalism and secularism. In other accounts, this worldview triumphed over more holistic and ecological conceptions of coexistence and communion with nature.[2]

Scholars have interrogated and undermined traditional conceptions of modernity. Instead of offering a master narrative of conquest, secularization, or disenchantment, several scholars now recognize that modernity is best understood by the tensions and complexities between competing ideas. The unresolved contradictions reinforced and reflected a modernity that was "Janus-faced," to quote Saler. Science and technology played key roles in this realm of competing representations and contradictory discourse. As historian David Nye and others have noted, technological marvels could provoke religious emotions, despite being the hallmarks of humankind's conquest of nature and enlightenment from a mystical past. In fact, certain technologies could evoke sublime emotions and transcendental longings, as modernity became increasingly complex and even contradictory.[3]

This reconciliation between extremes became widespread, particularly in Germany, where Willy Ley spent his early years. Given long traditions of *Naturphilosophie* that valued the role of speculation and wonder, many Germans embraced Romantic science as well as Romantic literature and arts. This appreciation for wonder, awe, and the complexities of nature reflected an attitude or sensibility that circulated beyond the laboratory or research center. Romantic science and popular science often became indistinguishable. Both realms became far more inclusive for amateurs, female naturalists, and imaginative thinkers.[4]

Romantic science found a broader audience, not only in Europe but also in the United States. When scholars investigated the cross-cultural currents in more depth, the story became much broader and transnational. For example, historian Aaron Sachs examined the popularity of naturalist/artist Alexander von Humboldt in the United States. Sachs argued that Humboldt offered a powerful alternative. Americans, like their German counterparts, appreciated Humboldt's "deep feeling of awe and appreciation for the great variety of landscapes and cultures." In this perspective, Humboldt offered readers a romantic naturalism and an ecological awareness, in which "Nature offered not only deep insights but also solace and sanctuary; the very image of a wild and overgrown landscape could move people spiritually, was perhaps even more valuable in times of need than religion." Humboldt's works offered a daring, interdisciplinary mix of science, art, wonder, and poetry. He inspired readers to worship "the cosmos itself, the beautiful whole that could not exist without each of its parts."[5]

In Sachs's perspective, Humboldt's ecological worldview enchanted many Americans before it evaporated at the turn of the century. The modernists (or positivists) who followed showed little reverence for nature or the environment. Likewise, they did not tolerate romance, wonder, or speculation. By the time of the Great War the Humboldtian explorer was a living fossil. Enthusiasm for imperialist adventures and triumphant exploration waned. This perspective fits with other narratives of disillusionment, due to the carnage of the First World War.[6]

Yet if the works of Willy Ley are indicative of a larger trend, the Humboldtian cosmos survived. Modern romantics, like Ley and Sagan, celebrated the beauty and power of Nature (often with a capital N) as well as the interdisciplinary combinations of science and art. They glorified Nature, even as they drew modernist designs for conquest and exploitation. These scientific intellectuals and showmen sought to reconcile extremes, while they kept the door open so that others could follow. They also glorified fieldwork over laboratory manipulations of

Nature. They saw themselves as heirs to an older tradition of ecological, alternative, and spiritual science and science writing. They promoted a spiritual vision of the cosmos, both above and within all human beings.

By following the life and career of Willy Ley, this book explores the spiritual dimensions of modern science, while interrogating traditional narratives of disenchantment. Ley's works embodied a Janus-faced modernity, with its many complexities and competing representations. Ley also represented a type of modern romantic, who celebrated wonder, awe, and the technological sublime. Simultaneously Ley voiced modernist faiths, while he relentlessly debunked pseudoscience and mystical thinking. He waged both a war for enchantment and a war for disenchantment. He reconciled these extremes by promoting a modern, scientific form of enchantment, while dismissing a (perceived) medieval, superstitious form of enchantment.

Additionally, Willy Ley's books and articles reveal much about popular science, whether in the context of Weimar Germany or postwar United States. These are understudied areas compared to scholarship on Great Britain. British historiography (particularly on the nineteenth century) has charted new territory by expanding its focus from scientific elites who popularized "science for all" to the ways in which science circulated outside institutions, journals, and laboratories. Consequently, we have a far more complex survey of the cast of characters, the variety of media, the many sites of communication, and the role of the public.[7]

As historians moved from the study of elite popularizers to the sites of circulation, they began to notice popularizers who did not fit neatly into the category of professional scientists. Opportunities arose for non-professionals who could communicate the broader significance of discoveries. Historians have begun to appreciate the intermediaries who occupied a public space between scientific elites and ordinary citizens. They have also acknowledged how popular culture could actively produce its own indigenous science or appropriate the claims and findings of elite practitioners. A larger cast of characters specialized in communication. Historians have begun to appreciate this broader variety of individuals and occupations, from female naturalists to journalists. Other scholars have moved into the twentieth century, which invites revisions of master narratives.[8]

In other national contexts, some progress has been made. Historian Andreas Daum has greatly enriched our understanding of nineteenth-century German popular science (which influenced Ley's perspective and writing style). Daum's survey of writers, clubs, organizations, and publications lays the groundwork for future studies. Likewise, other historians have taken readers into the zoos, classrooms, and natural history museums of nineteenth-century Germany. We have a richer understanding of the nexus of media, popular science, and consumer culture. Germany lagged somewhat behind Great Britain in mass markets and cheap print. Yet a growing market welcomed educational texts, particularly in Berlin. Ley grew up in this cosmopolitan hub, which offered science for all, as presented by science writers and popularizers. Despite their status as outsiders or popularizers, the science writers acquired public reputations as trusted authorities on a wide range of subjects, especially natural history and astronomy. Additionally, their styles of popularization often combined science, art, and poetry. Compared to British popular science, the German variants may be far more romantic as a genre. One could also connect the wide circulation of natural history texts to various back-to-nature movements that stretched from *Wandervogel* activities to the survival of German popular romanticism.[9]

When compared to British and German historiography on popular science, American historiography has struggled to develop and shed its baggage. In traditional narratives, the popularizers first emerged in the nineteenth century to educate the public, yet they engaged in a losing battle, particularly when journalists, publicists, and advertisers controlled mass media. Accordingly, twentieth-century popular science transformed into sensational claims and dubious "infotainment." The public received superstition, sound bites, and quack cures. Capitalism undermined the endeavor. In this perspective, superstition and pseudoscience won the battle for the public's attention. "Real science" was dramatized and marginalized. Media and its producers undermined the agenda of science educators and popularizers.[10]

Most scholars focus on the allegedly legitimate scientists who struggled to adapt to a changing media environment. There is an explicit difference between scientists and cultural producers. The scientists either reluctantly cooperated, compromised, or resisted in various media where sensationalism sold. They offered truth to an audience craving awe and wonder. They struggled to entertain without sacrificing the dignity and truth of their science. They lost the battle. Popular science, in the end, became compromised science, anti-science, or pseudoscience. Capitalism transformed it from a pure essence to a vulgarized titillation.

This biography adopts a different perspective. The life and works of Ley are indicative of the rise of scientific journalism and the media-savvy generalist. Ley occupies a place within a network of scientific intellectuals who embraced mass media and public education. The science writers did not retreat in the face of superstition or sensationalism. They engaged with the public, while they contrasted themselves to isolated and ineffective specialists. They promoted a type of romantic science and a form of the "scientific spirit" that democratized science for all. They understood the connections, the commonalities, and the whole in a way that catered not only for public consumption but also for public participation through forums, clubs, lectures, newspapers, magazines, books, and broadcasts. They saw the specialists as lost in isolation. As generalists, they produced intelligible books and articles.[11]

Therefore this biography is not a story of a scientist versus media. It is a story of a scientific intellectual who succeeded, first in Berlin during the 1920s and then in New York City after 1935. Ley did not approach a microphone reluctantly. As a science writer, he democratized knowledge while celebrating the audience's preference for wonder, awe, and mystery. Entertainment kept a people's science alive. Entertainment allowed space for an alternative, romantic science. Both entertainment and science won. His victories demonstrate how popular science flourished in the realm of mass media. Ley's career provides a window into a world of writers, publishers, and producers who democratized knowledge.[12]

As a science writer, Ley should not be labeled simply as a translator, simplifier, or (some would say) vulgarizer. His skills as a generalist were far more valuable than a specialist's expertise. His ability to communicate set him apart. Like other romantics, he celebrated the links between American democracy and amateur science, in which everyone could participate by learning science while enjoying the enchantment. He intentionally blurred the distinctions between science communication, public engagement, and scientific thinking. He encouraged his audience to experience the wonders of nature and the marvels of science. His works embodied science in the vernacular. He shared the stage with a large camp of scientific intellectuals who provided syntheses as storytellers who inspired audiences. Ley's life and works reveal popular science on its own terms, as it flourished and circulated in informal environments.[13]

Additionally, Ley's books and perspectives further situate popular science within traditions of American anti-authoritarianism. As other historians have noted, many Americans resisted or resented hierarchies of knowledge. Instead, republicanism demanded participatory engagement, along with a democratization of knowledge, skills, and arts. Media scholars have long noted this characteristic of popular culture. They have explored the egalitarian, rowdy, and volatile world of nineteenth-century theater. More recently, scholar Richard Butsch has documented "the citizen audience" to show how the very idea of citizenry found its most poignant expression in public participation at spectator events. Americans often expressed identity in anti-authoritarian terms that distrusted elite expertise. Americans preferred showmen, like P. T. Barnum or Buffalo Bill.[14]

As a famous rocket expert, Ley became a showman of the Space Age. This identity is an important part of the story. Arguably, Ley's space-related media blended genres while offering an uplifting and romantic vision of science, technology, and the ascent of humankind. Ley's role in the history of spaceflight demonstrates the importance of thinking about the circulation of texts and the role of individuals who operated outside scientific laboratories and machine shops. One could even argue that Ley is an exemplary figure who recognized that science is a form of communicative action, whereby "knowledge in transit" puts the popular science writer at the center of the scientific enterprise. Considering that Ley did so much to inspire Americans to support a publicly funded "conquest of space," the scientific and technological accomplishments of the era were, in some ways, a consequence of media.[15]

Willy Ley was the most important publicist of the American Space Age. Metaphorically speaking, Ley was the one of the people behind the curtain, pulling levers and adjusting lights. Or he took center stage in front of audiences, cameras, and microphones. In the 1950s and 1960s he shared the stage with famous engineer Wernher von Braun, who has been called "the single most important promoter of America's space effort in the 1950s and 1960s." This biography challenges von Braun's status as a preeminent popularizer. Readers may feel bewildered by both the cottage industry of books on engineers and astronauts and the lack of books about the movement's chief publicist. The present work also takes readers further away from the traditional and institutional boundaries of space history. A biography of Ley provides a window into the vast forest of media, publishing, and publicity.[16]

Ley is relevant for historians of spaceflight for another reason. He became their founding father. His historical book, *Rockets, Missiles, and Space Travel*, went through twenty-one printings from 1944 to 1969. As noted by other scholars, this book became both a primer in the history of spaceflight and a textbook in the popularization of rocket technology. Not only was it the first historical grand narrative; it also educated American audiences about rockets and space travel. To a large extent Ley's memoir and even his personal perceptions of key individuals have deeply influenced secondary literature on the history of spaceflight. Ley's depictions of key visionaries and their subordinate inventions privileged the role of individual genius over the role of states or even institutions. His historical synthesis of "Prophets with Some Honor," with their corresponding "steps" in the right direction had a lasting impact on the field.[17]

Although Ley presented his narrative in international terms, he simultaneously privileged the role of the German pioneers. Historians have often struggled against the power of his narrative to establish a place for indigenous American rocketry. One could read the initial turn toward institutional histories or biographies of astronauts as a struggle to counter-balance Ley's German-centric accounts. Yet even when struggling against Ley's perspectives, historians have been deeply influenced by his claims. We are in fact still untangling many of the facts from the fabrications. Ley's narratives contained much misinformation, due to his reliance on ex-Nazi engineers and their selective memories of the underground factories that utilized concentration camp labor to construct V-2 rockets. Thus many historians of spaceflight have long struggled to uncover the truth behind the stories told to Ley, which he repeated for a mass audience. In other examples, Ley's recollections could be self-serving, as memoirs naturally are. In this biography I recognize and untangle the power of Ley's narratives on the field of space history.

In spite of this necessary focus on space history, I argue that Ley's roles as a spaceflight advocate and historian were smaller parts of a broader crusade to educate the public, conflate science and imagination, and promote a romantic appreciation of nature. In fact, the key to understanding his many works on spaceflight is to appreciate the blending of genres and the overlapping interests that united his fascination with rockets with his love of natural history. It is extremely important to explore his many adventures as a romantic naturalist. The conquest of space related to the broader conquest of nature. It was the culmination of a long history of exploration, wonder, and discovery. Ley is a prime example of a popular science writer who continued in a long and transnational tradition of exploring the wonders of the world. He united natural history and popular astronomy together in a literature for mass consumption. He encouraged a romance with nature. The way in which historians describe the tropes of spaceflight literature can be applied to all of Ley's works, in which he voiced a spiritual quest for new worlds and final frontiers.[18]

Ley also sought to debunk (what he viewed as) nonsense, propaganda, mystical thinking, and pseudoscience. His broader crusade as a public educator had a forceful side. Here Ley resembles other public intellectuals, writers, and historians with less self-consciously romantic ideals and aspirations. From scientific historians like George Sarton to science writers like Isaac Asimov, these intellectuals strove to dispel nonsense, discredit pseudoscience, and purge irrationalism from the ranks of the popular. They prescribed a heavy dose of scientific thinking as an antidote to superstition, mystical thinking, and cultlike obedience to a premodern mindset. They also engaged in a related political struggle, as they equated both fascism and communism with a generalized totalitarianism.

The science writers' view of the world was stark, given the contexts of the Second World War and the Cold War. In one corner stood American democracy and science, both of which promoted freethinking and the open search for truth. They claimed that both democracy and science encouraged a critical anti-authoritarianism and a republican ethos. Not coincidentally, this camp celebrated the champions of science as revolutionary and sometimes tragic figures who spoke truth to power. The modern scientist, they argued, had the same social responsibility to save souls from the spread of totalitarianism by preaching the scientific spirit, with its instinctive disregard for dogma. In the other corner stood totalitarianism, spreading as a powerful menace to threaten all political and scientific progress by eliminating free inquiry and speech. Totalitarian or medieval pseudoscience functioned by cultlike adherence to authoritarian truths. The totalitarians did not tolerate dissent or critical thinking. Consequently, real scientific breakthroughs could not happen in a closed society.[19]

This outlook makes historians cringe. To make a distinction between real, democratic science and fake, totalitarian science is an absurd and outdated project. Readers should also wince at the outdated distinctions made between astrology and astronomy, alchemy and chemistry, and eugenics and biology. These distinctions have long been discredited by historians of science, who do not easily demarcate the real from the fake, as the science writers did. Ley went a step further, by offering grand narratives about the "Dark Ages" and proto-totalitarian mentality. Academic historians have discredited these terms and associations since the 1930s. Nevertheless, the language circulated in popular culture. These perspectives teach us much about public discourse during times of war.

In spite of this crusade against cultist thinking, these intellectuals shared a mystical and even deeply spiritual outlook. As becomes clear from many quotes, they often invoked religious language while speaking about their faith in science and technology. For example, one sociologist asked if science could ultimately save humanity. He answered, "When we give our undivided faith to science, we shall possess a faith more worthy of allegiance than many we vainly have followed in the past." This view of science could be labeled as deeply mystical and enchanted. Even as these intellectuals participated in the various pseudoscience wars that flourished in mass media, they did not promote a completely disenchanted science, void of transcendental qualities. Rather, they promoted a romantic and spiritual scientism that cherished communion with a higher power, the cosmos.[20]

Chapters 1 and 2 provide essential background for many of these arguments by focusing on Ley's formative years in Germany. These chapters illustrate how popular science, romanticism, and science fiction influenced Ley. They also demonstrate how he embraced amateur clubs, popular media, and the role of an intermediary who could translate complex concepts for a general audience. These activities and media defined Ley's entrance into the world of science. Chapter 2, in particular, illustrates how Ley constructed his identity as a freelance writer in reaction to several perceived "others": the isolated and ineffective scientist and the effective but dangerous pseudoscientist. Chapter 3 focuses on his perceptions of the decline of amateur groups and rocketry, due to the rise of totalitarianism and anti-science in Nazi Germany. In this chapter I argue that his perceptions of real science were constructed in reaction to the perceived spread of irrationalism that transformed Germany from a scientific democracy into a totalitarian state. Chapters 4 and 5 recount Ley's early years in the United States, after he fled Nazi Germany in 1935. Whereas chapter 4 explores his early adventures as a freelance writer in New York City, chapter 5 documents his rise as a public educator during the Second World War. These accounts begin to put forth a more complex argument surrounding the role of science writers and popular science, in the context of war and a perceived totalitarian menace. Chapters 6 and 7 then turn to the postwar years, when Ley established himself as America's top expert on rockets, missiles, and space travel. These chapters highlight the nexus of popular science and mass media, along with the various crusades to educate, entertain, and debunk. Modern enchantment is situated within the somewhat contradictory pseudoscience wars that flourished in popular culture. Chapters 8 and 9 follow these trends throughout the late 1950s and most of the 1960s, when the Space Age exploded in popular media. These accounts illustrate Ley's overarching tactics as a popularizer of awe, wonder, and the technological sublime. Simultaneously they present his efforts as a popular science writer in relation to a larger camp, including historians of science, whose actions and tactics were quite political. Lastly, chapter 9 documents Ley's declining prestige as a freelance historian of science. As the history of science transitioned from an open and cosmopolitan scene into a more isolated and academic field, Ley was one of many scientists-turned-historians who were excluded and ostracized. Younger historians called his perspectives outdated and self-serving. They viewed his style of popular writing as old-fashioned. They used many jargon-filled labels of scientism, Sartonism, and modernism to discredit an older guard. Arguably, there is a larger story here about the academic institutionalization of the history of science during the 1960s that invites readers to ask, "What was lost?" Those readers might experience some degree of nostalgia for a time when academics and popularizers mixed ranks.

Youthful Horizons

In a later autobiographical note to readers, Willy Ley recalled a defining event of his childhood in Berlin, circa 1917. Standing before his teacher and peers, he had been given the task of an essay assignment and class presentation on the question of "What Do I Want to Be When I Am Grown and Why?" The question could not have been an easy one for many of his classmates. Some had probably lost their fathers in the ongoing Great War, which killed (on average) 1,300 Germans per day. This loss of life eventually created over 350,000 widows and left over 730,000 children fatherless. Other students may have been more fortunate, because their fathers survived. In the postwar years these children-turned-young-adults witnessed the familial effects of war neurosis and shell shock as veterans struggled to readjust to civilian life. This generation of fathers was defined by trauma, death, and modern, industrialized warfare.[1]

The children experienced the trauma, death, and indifference of the home front. They watched their mothers struggle to provide sustenance amid bread and potato shortages, food riots, and strikes. They endured a blockade, which indirectly killed a million people during the war. The everyday realities of wartime urban life had also, in the words of historian Belinda J. Davis, "shattered the illusion of upholding the ideal family and the role of its members." Life centered less on an ideal nuclear family and a stable home. Instead, families experienced hardships in public streets. Poorer women struggled most, as they fought for potatoes after waiting in long lines. In this harsh battle for the basic necessities of life, these women and their families barely survived.[2]

Young Willy had an easier childhood than some of his classmates. He was born Willy Otto Oskar Ley on October 2, 1906, in Berlin. He was the son of Julius Otto Ley, a traveling wine merchant, and Frida (May) Ley, the daughter of a Lutheran church official. Ley saw little of his parents after age seven. In a 1955 autobiographical note, Ley recalled, "It so happened that my parents were in London when that war broke out. I was in Berlin all the time, living with relatives." British authorities interned Julius on the Isle of Man until the end of the war. They allowed Frida to return to Germany, carrying her newborn baby Hildegaard. According to some accounts, she did not remain in Berlin. After leaving the newborn with her sisters, she may have left to work as a milliner in a different city. Ley's aunts took care of him and his sibling, with the support of other relatives tied to the German Lutheran Church.[3]

These were the two sides of Ley's background: business and the church. These two sides represented his safe career paths. However, he had little regard for family traditions. He later described his family in uninspiring ways: "A possible future biographer will have a hard time finding family background for either the scientific or the literary side of my inclinations and activities." Unlike his family and many of his peers, Ley had a creative spirit. As a gifted student, he questioned his teachers. He eagerly read morning newspapers. Ley was intellectually curious and often self-taught. For example, his Realschule may not have required traditional courses, such as Latin and Greek classics. Nevertheless, Ley learned Latin and read many of the Greek classics. His fascination with Greek mythology lasted a lifetime. Ley loved reading the books in their original languages. He remembered: "Like every future author or scientist I ever heard of, I was an omnivorous reader, first in German only and then, as schooling progressed, in Latin, French, and English too." It is tempting to imagine Ley as a teenager, absorbing himself in Latin texts, as if they offered overlooked gems, waiting to be discovered by someone who did not simply rely on translations or later books. This distrust of secondary sources also lasted a lifetime. In Ley's perspective, a good student not only questioned his teachers but also investigated the sources. If the evidence did not support old myths, legends, or superstitions, an intelligent student questioned antiquated beliefs. Often a freethinker destroyed bad systems, while speaking truth to power.[4]

The Heroic Scientist

Given Ley's perspective on learning, a certain genre of literature excited his imagination: tales of exploration, both true and imaginary. He enjoyed books that glorified fearless explorers who discovered amazing things. He consumed books that narrated marvelous quests, mysterious places, ancient secrets, and hidden worlds of wonder. For something to be an interesting story in either science or fiction, it usually included something "amazing," "astounding," or "marvelous." Explorer travelogues probably occupied the same shelf as futuristic tales of imagination.

By far his favorite writer of fiction was scientist/philosopher Kurd Lasswitz whose book, *Auf Zwei Planeten* (*On Two Planets*, 1897), occupied a special place in Ley's collection. As a literary fantasy about first contact with intelligent Martians, the book made a lasting impression on Ley. He even described it as "one of the best and most interesting novels of German literature." The novel's critique of European imperialism did not impress Ley. He summarized the morality tale crudely: "It was . . . basic psychology to show that the highly ethical Martians, when confronted with terrestrial stubbornness, quickly revert to war." However, the author's "solution to the problem of space travel" impressed Ley greatly. The novel included Lasswitz's mathematical calculations of trajectories, orbits, and rocket launches. Additionally, Ley admired the book's representation of scientists as explorers who fearlessly sought out the unknown. The book glorified scientific adventures, new discoveries, and breathtaking landscapes.[5]

Ley also read the works of nineteenth-century author Jules Verne. In Ley's judgment, Verne's novels "expressed confidence in the powers of science and discovery, a confidence well fortified." Verne's adventures were also "romantic," since they recounted "explorations of the unknown." Men of science crossed political boundaries to traverse the earth in the air. These adventurers also discovered especially weird things in the depths of the ocean. Rarely did they get wrapped up in political limitations. Instead, politics got out of the way in favor of marvelous inventions, wondrous machines, and daring adventurers. Fearless scientists journeyed to the center of the earth and the surface of the moon. Exploration and science were identical quests.[6]

Through Verne's novels Ley consumed an image of the "scientist as adventurer." Scholar Roselyn Haynes described the common tropes surrounding the modern version of a traditional Romantic hero, "now allied with science rather than opposed to it." These characters served as "humanity's advance guard," by expanding the frontiers of both space and time and "transcending mankind's former limitations." These "technological knights" boldly expressed their right to dominate nature, the universe, or "whatever alien societies they encountered." Most likely, Ley dreamed of being among the ranks of these new explorers, who daringly crossed frontiers.[7]

In these fictional stories the scientist was a swashbuckling adventurer who embodied fearlessness in the quest for new worlds. Additionally, the explorer expressed anti-authoritarian tendencies. The scientist interrogated and tested conventional knowledge. He doubted the reliability of established thought. He would hatchet his way through thick jungle bush to discover the truth about the world. For Ley, Verne's heroes exemplified "a new attitude." Ley wrote: "Consistently his heroes . . . do things for themselves. They do them in a novel way. They don't do things in a traditional and poor and inefficient manner for the sake of tradition. Nor do they look for 'lost arts.'" He continued:

> Instead of yielding to the traditional modesty of being "insignificant sons of great ancestors," they act with the full knowledge that their time has surpassed any preceding time. They know that they know more than their ancestors. . . . They don't hesitate to cruise under the seas or fly through the air. And to them the problem of reaching the moon is what it really is: a question of attaining a sufficiently large velocity in the right direction at the proper time.[8]

According to Ley, Jules Verne's extraordinary voyages stimulated the imaginations of young readers and interested them in "the connection between the past and the future, between the real and the possible." Verne also expressed a deep fascination with the United States as a land of daring explorers and brave engineers. As noted by a literary scholar, American society was often portrayed as "one in which scientific and technical problems are of concern to the man on the street corner. They belong to the people, rather than being set apart as they are in the Old World, in the dusty studies of the Academies and scholarly societies." The United States possessed a great scientific and industrial frontier, which was open and democratic.[9]

Verne's stories impressed Ley. He consumed representations of scientists as bold explorers, who bravely set out to face the unknown and conquer new frontiers. New explorers would act accordingly. Nature would yield its spoils. The scientists would penetrate its secret realms to dominate, reorder, and repurpose the spoils for the benefit of humankind. The riches of the frontier would be marvelous. Most likely, as Ley and his family struggled for basic sustenance like bread and potatoes, he dreamed of those distant frontiers.

The Heroic Engineer

Many historians have argued that the First World War served as a "technological maelstrom" that diminished or destroyed hopes and dreams. Machine guns, tanks, mortars, poison gas, and other innovations of the era transformed war from an honorable and noble enterprise into an assembly line of human carnage. In this regard, the Great War facilitated a crisis of Western Civilization as well as a crisis of masculinity. Scholar Michael Adas argued: "Little that was glorious or noble could be found cowering in ditches in the midst of a wasteland glutted with the bloated bodies of dead men and animals." In the perspective of famous German soldier Ernst Jünger, science and technology had converged to create "a cosmic, soulless force before which man almost disappeared." Historians have analyzed similar perceptions of the loss of chivalry due to industrialized warfare. Accordingly, the Great War was a watershed moment that caused many European intellectuals not only to question their faith in machines as the "measure of man" but also to reevaluate their very notions of civilization, progress, manliness, and chivalry. Consequently, according to many historians, public interest in science declined due to associations with poison gas and industrialized warfare.[10]

Ley and his fellow students viewed these events in less critical ways. Although they had endured the civilian effects of total war, there was no moment of great disillusionment with technology and science. It is doubtful that Ley or his fellow students thought in terms of the ideologies of Western dominance or the "measure of man." Instead, Ley belonged to a generation who would continue to celebrate technologies and other modern marvels, especially aircraft. In fact, Ley recalled, "One of my earliest memories is seeing one of the airships built by Count Ferdinand von Zeppelin circling over Berlin." Ley celebrated new heroes, embodied in the image of the aviator. As summarized by historians, the war ace became a symbol for the continuity of chivalry. He was a both a noble warrior and a skilled engineer. In the words of soldier Ernst Jünger, aviation represented "a fiery marriage of the spirit of ancient chivalry and the chilling bleakness of our forms of labor." The sky warriors retained control over their tools. Their skill and daring still mattered. Jünger further commented: "In them one finds the highest workerly and soldierly virtue stamped in fine metal, combined with intellect applied to the tasks in hand, and not without a certain freedom of style and an aristocratic delicacy." As described by historian Eric Leed, their aircraft enabled them to rise to an altitude "where, once again, war was a unified human project." Fire and steel joined forces in the conquest of the sky, as a new breed of heroic engineers took center stage.[11]

Ley recalled his adoration of the "Captain Future" stories, which narrated the adventures of pilot Captain Mors, who led thrilling adventures around the world and even into space. According to Ley, these stories were "outright science fiction that showed evidence of wide reading and even research on the part of their author." Ley particularly enjoyed the pilot's aerial adventures to Tibet as well as his attempts to divulge the secrets of Martian solar energy weapons and Venusian "heat-beams." Despite the odds, Captain Mors always prevailed, saving his ship by thinking quickly and flying swiftly. It is doubtful that Ley considered the ways in which his enthusiasm for Captain Mors related to a perceived loss of chivalry and masculinity. Like most Germans, young and old, he was simply fascinated with technology and optimistic about the future, in spite of the Great War. There was no watershed moment of grand disillusionment with science, technology, and Western values. Rather, a new era of exploration had dawned. Ley looked to the sky. He peered into the future.[12]

The Heroic Science Writer

In his own words, Ley "grew up, so to speak, in the shadow of the Museum of Natural History in Berlin." He explored every nook and cranny of this museum, which served as a scientific cathedral. Ley later expressed his surprise at the exciting discoveries made in less traveled hallways and exhibits. "I spent much time in wonder," he recalled. He discovered "especially weird things in it." "My first love," he remembered, "had been fossil animals." In fact, one of his fondest memories related to a special hall devoted to paleontology. Ley described the scene: "Almost all the way to the high ceiling it was 'paneled' with large slabs of red sandstone which was even more intensely red because of the sunlight that struck them slantwise through tall windows." He added, "One turned away from that wall of red sandstone with a sense of mystery." Soon his mind turned to astronomy, zoology, and botany before his interests focused on the paleo-sciences. He later explained: "The past periods of earth's geological history fascinated me." According to later publicity material: "He was, from his early high school days, fascinated not only in all aspects of scientific fact, but by the history behind scientific discoveries."[13]

His rambles through museums, observatories, and zoological gardens were complemented by his wandering discoveries in libraries. Judging by a later inventory of his library, Ley was an avid reader of popular science, and his bookshelf included popular astronomical books by Dr. M. Wilhelm Meyer, director of the Urania Observatory in Berlin. Ley later recalled, "One of the first books I ever bought—a mixture of curiosity and nostalgia—was a small volume by the German astronomer . . . called *World's End.*" His bookshelf also contained three volumes of Alexander von Humboldt's *Kosmos*. Other favorite authors likely included Camille Flammarion, Richard and Mary Proctor, and Percival Lowell. Ley may have read Kurd Lasswitz's nonfictional books, which explained the philosophy of Immanuel Kant. Lasswitz wrote additional magazine and newspaper articles, later reprinted. Lasswitz described "the profound and inextinguishable longing for better and more fortunate conditions" on other worlds.[14]

These writers did not bully readers with unnecessary jargon, non-sensical terminology, or long equations. Rather, they enchanted the material for a general audience. They served as interpreters and educators. They combined imagination and science. They engaged with the public. Many of these popularizers must have benefited from the same types of developments that occurred in Great Britain, as documented by historian Peter Bowler. Although the war and postwar economic chaos disrupted the publishing industry, Berlin was a cosmopolitan environment, where young adults like Ley consumed tales of fantastic discoveries, unsolved mysteries, and new frontiers. Ley read a new generation of popular science books, produced for a younger audience. These books provided intelligible and condensed versions of textbooks. They both educated and entertained readers. Simultaneously, they encouraged readers to think for themselves, rather than simply trust experts.[15]

Ley's favorite science writer used the penname of Dr. Theodor Zell (Dr. Cell). Dr. Cell was a scientific celebrity, capable of educating and entertaining a mass audience. Ley recalled, "His specialty was to explain actions of animals which seem mysterious or senseless to the casual observer." Although Zell never experimented with animals, his observations and writings made him one of the most skilled debunkers of zoological myths. Ley greatly admired this ability to debunk false claims, particularly those fables and myths that circulated widely without being scrutinized. Ley appreciated the role of a popular science writer when it came to setting the record straight. Dr. Cell did not need to conduct elaborate experiments in an isolated laboratory. He simply had to observe nature. He also uncovered scientific truths by exploring history. He combined history and science. Overall, he taught ordinary people about science and exploration.[16]

Additionally, Ley enjoyed the works of naturalist and science writer Wilhelm Bölsche. He probably adored Bölsche's *Love-Life in Nature* (1898) and *The Victory of Life* (1905). Bölsche wrote poetically about science, while using illustrations and paintings to convey the mysteries of the planet and its strange creatures. Bölsche combined science, poetry, artistic imagination, and literary prose. In the preface to a new edition of his most famous book, *Love-Life in Nature*, Bölsche wrote: "My book is addressed to all rational people who have the courage to form a philosophy of life for themselves." He then offered no apologies for the tone of the book, which combined personal reflections, philosophical ruminations, and scientific theories. He explained: "The bridge connecting the field of the strictly scientific . . . with the world of sovereign thought, which seeks the whole, leads across art, art with all its instruments, even humour." Bölsche rejected the need for a "special solemnity of tone," when it came to presenting scientific and philosophical material for lay audiences. He argued, "An artificial assumption of dignity is an absurdity when pure, genuine human beings get together."[17]

The book took readers directly to the sites of wonder, discovery, and awe, while the author spoke to those readers:

> I should like to discuss many things with you. . . . But look out into the boundless brilliance of the sea. . . . Look into the firmament above and behold its infinite dazzling purity. Out of this blue of eternal space the worlds rained down like silver dust. How many alarming, horrible things the depths of this flood concealed, and still conceal. And yet, on the whole, it is a wondrous blue, into which the soul dives as into a bath of peace.[18]

The book emphasized the spiritual and scientific connections to a cosmic whole. "You are on earth and the stars are above you," he argued. "In the widest sense, you are a cosmic body as they are. Size matters nothing. . . . You and Sirius both of you swim in the fine cosmic substance which the physicist calls the ether, like two fish in the same vast pond."[19]

Bölsche blended genres as he explored the fringes of zoology, encouraging his readers to become fellow naturalists and freethinkers. He spread the message of humanistic naturalism. As part of this crusade to bring science to the people, Bölsche promoted a deep connection between the scientific spirit and the poetic imagination of human beings. He hoped that a love of science would provide the means of upward mobility, which is why he helped to create Germany's first "peoples' school."

Willy Ley admired and respected Bölsche. He shared many of the same values and perspectives. He learned to see the world in terms of wonder. He marveled at the mysteries. He looked to the future with an enthusiasm for science and technology. He believed that he lived in a new age of scientific discovery. New machines would be the measure of humankind. Science was open to all.

The Heroic Explorer

Although these perspectives crystallized during his teenage years, the seed had been planted back in 1917, when he first answered his teacher's question about his life ambitions. Ley announced: "I want to be an explorer." The reaction of his teacher, he recalled, was patronizing and unimaginative: "My teacher made a little speech, saying that I deserved a good mark for my style and that the reasoning, 'such as it was,' was logical too. Except that the whole thing was, of course, nonsense. A boy with a family background of business on one side and church on the other just doesn't want to be an explorer, or, if he does, he certainly won't become one." The speech left him unconvinced. He recalled: "I kept exploring, in a manner of speaking, looking especially into such corners as others had neglected." Several years later, his path was still somewhat murky, yet his ambitions were clear. He remembered, "By the time I was ready to graduate from high school I was sure that I would become a geologist." He also remembered, "To tell the truth: the border lines of those sciences interested me more than the actual material. I found the history of zoology more fascinating than zoology itself." "I didn't know what I wanted to be," he also admitted, adding, "I read omnivorously, and my interests turned to science."[20]

Near the completion of his primary schooling, Ley moved to Konigsberg, where his father now operated a liquor business. Ley may have moved into his father's house. Only one fact about his time in Konigsberg can be verified. Ley became deeply fascinated by the "local phenomenon" of amber, which made the Baltic coast famous. He spent much time researching the myths, legends, and known scientific facts about its origins. It was a geological mystery that demanded a resolution. Thus Konigsberg was the perfect site for an aspiring young geologist, fossil hunter, and gem digger. Ley began to study (informally) at the University of Konigsberg. He was confident of his future potential in the realm of science. He was more determined to embark upon a "lifetime of interest, a lifetime of collecting material, a lifetime of 'exploring.'"[21]

Unfortunately, the year was 1923. Ley's dreams had to confront the political and economic realities of a world turned upside down.

From the Earth to the Moon, via Berlin

When Ley turned seventeen in October 1923, he sought to embark on a scientific career. Unfortunately, widespread financial and political instability dashed his hopes. Due to massive postwar inflation, money had become almost worthless, while confidence in the new republic dwindled. By late summer 1923 the value of 1 million German marks equaled the value of 1 American dollar. Ley later recalled the everyday realities of the postwar situation, when he paid 30,000 or 50,000 marks to ride a street car. It was not an encouraging scene for an aspiring student.[1]

Ley's family had a difficult time. Father Julius Ley's liquor business failed, due indirectly to inflation. This failure eliminated any possibility of financial support for his son's educational ambitions. By early 1924 Willy Ley had moved back to Berlin to find work. Some biographical articles state that he became a full-time student during these years. For example, writer Sam Moskowitz claimed that Ley entered the University of Berlin, taking courses in anatomy, zoology, and astronomy. Ley probably attended public lectures, not courses. Additionally, his studies were intermittent. "I was a young man," he later admitted, "wrestling a living from a kind of permanent economic depression and studying zoology, some paleontology and a little astronomy at night."

He recalled, "To stay in school I had to work, and became a bookkeeper in a bank, attending school evenings." Ley's situation was far from unique, as the percentage of working students rose dramatically in the early 1920s.[2]

Ley educated himself outside classrooms and institutions. He later mentioned that he had little patience for expert lecturers who demanded a passive audience. Instead, he explored libraries, museums, and zoological gardens, always with an eye out for oddities. In this sense Ley occupied the non-academic space of an emerging dichotomy between natural history and professional sciences. As the print market expanded, he read popular and educational books. He also explored places that blended scientific curiosities, the political values of the middle class, and the experiential appreciation of nature. Consequently Ley adopted neo-Humboldtian views. He appreciated the nineteenth-century explorer's attempt to make connections and discover the "wholeness" of science. Additionally, Humboldt had blended nature and art, along with science and poetry. Ley's favorite writers had followed in Humboldt's footsteps. They harmonized various disciplines into a unified whole. They embraced enchantment, wonder, and spiritual connections to nature.[3]

These writers followed in the footsteps of earlier vitalists and proto-environmentalists, who argued that scientists could not understand how a part functioned without understanding the larger, interdependent web of life. This anti-mechanistic discourse also overlapped with political and cultural debates. By the Weimar years some of this rhetoric grew ugly and reactionary, particularly in the association of Jewish scientists with an allegedly abstract, soulless, or materialistic science. There are no indications that Ley thought along these terms, although he may have, given a detour into fascist politics in 1925. Most likely the young Ley simply loved science and exploration. He loved the way holism mixed genres. He searched for a meaningful communion with nature. He also continued to ruminate about the potential frontiers of the future.[4]

Window-Shopping for a World of Tomorrow

On a cold October day in 1925 Ley took a walk down Berlin's Friedrich-straße, an area famous for its storefronts and tourist attractions. He had just turned nineteen years old. Berlin continued to recover from the postwar social, economic, and political chaos. Since the election of Hindenburg, the streets had become safer. Due to economic stabilization, some Berliners had disposable income again. Food was no longer a scarcity or barter currency. Friedrichstraße buzzed as consumers and commuters familiarized themselves with new commodities and advertisements, often covered by vibrant and futuristic styles. In years to come Friedrichstraße would be called a center of life rather than a simple transportation hub.[5]

Not only had the Weimar scene liberated many of the modernist movements, but also the scene grew increasingly international and cosmopolitan in style. Grand experiments occurred, from the political trial of the republic to the flourishing of expressionist art, quantum physics, and Freudian psychoanalysis. As noted by an influential scholar, many cultural rebels sought answers to questions, and they often "turned to whatever help they could find, wherever they could find it." They displayed little respect for frontiers or boundaries. Their expressionism and their Americanism left marks upon the Berlin landscape. The city buzzed with strange ideas, foreign imports, passionate politics, and international theories. The media landscape promoted unconventional ideas and representations of class and gender.[6]

The city also buzzed with new machines. Not only did automobiles and public transportation continue to revolutionize how Berliners commuted, but also mass production and scientific management revolutionized how they worked and what they consumed. This explosion of new technologies and techniques has led some historians to generalize about the public fascination with new technology. According to historian Richard Vinen, the postwar era witnessed the birth of a culture that became obsessed with machines. Vinen argued, "In no other period of the twentieth century did educated Europeans talk so much about the impact of science on their lives." This perspective undermines narratives of disillusionment.[7]

Much excitement focused on the technologies of flight. The popularity of amateur gliding increased. Additionally, 1925 marked the creation of the airline Lufthansa, after Allied restrictions on civil aviation lessened. Transoceanic flights continued to dazzle crowds. Germans would gather to witness aerial spectacles or a passing Zeppelin. For many historians, much of this public enthusiasm can be linked to German nationalism, because aviation forecast the revival of a powerful empire. In this view, an expression of aerial enthusiasm involved a deeply political act. For example, historian Peter Fritzsche argued, "Machine dreams mingled with national dreams. . . . Germany appeared to hold its own, despite political and economic hardships. Between the two World Wars, it was aviation [that] took the measure of progress."[8]

Ley would have agreed with the last sentence. Technologies of flight were technologies of the future. They fascinated him deeply. Like other Germans, Ley probably took every opportunity to marvel at airplanes and airships, which in 1925 were still a sight to behold. Perhaps as he walked the streets of Berlin he paused and looked up every time something buzzed. So too did fellow pedestrians who for the first time were seeing airplanes outside the pages of newspapers or magazines. Yet, unlike many Germans who may have embodied Fritzsche's description of nationalism and "airmindedness," little evidence suggests that Ley fantasized about the revival of a German empire. Aircraft represented the conquest of the sky. Like other Germans, Ley looked to the future. Machines still represented the pinnacle of human progress.

As Ley strolled down Friedrichstraße in 1925, his direction was clear. Economic and political stabilization had brought a sense of security. His studies slowly progressed. He informally trained himself to become a practicing scientist. His interest in geology had grown stronger. He later recalled, "A young man, of course, never knows what he's going to be and he is never as definite as his answers indicate. But it was geology, then." He had little ambition to become a writer, despite his love of popular science books. Nevertheless, he received a surprising check in the mail, rewarding him for correcting a scientific inaccuracy in newsprint. Ley had a few marks to spare, as he window-shopped for a good book.[9]

Other Berliners had long engaged in this type of window-shopping. Now new and exciting displays altered the storefronts as commodities dazzled under electric lights. As with much of the German experience with modernity, these surfaces and spaces could be shocking. By 1925 the urban scene became dazzling and dizzying, full of wonder, shock, and awe. As Ley roamed the streets of Berlin, he most likely marveled at the advertising, lights, and consumer products. As described by scholar Andreas Killen, Berlin was a city of "hypermodern urbanity . . . tinged with fever, a hothouse of unreality."[10]

Suddenly Ley saw an object that piqued his interest. He remembered: "I paused at a bookstore window to see on display (translated from German): *The Advance into Space—A Technological Possibility*, by Max Valier, a writer of whom I had heard vaguely." The cover of the book depicted a spacecraft en route to Saturn. "As far as I am concerned," Ley recalled, "the Space Age began."[11]

The Investigation of Sources

The exposure to Valier's book was Ley's first encounter with a contemporary popularizer of spaceflight. Through his writings and publicity, Valier had achieved some degree of fame. Ley soon learned that Valier was a self-proclaimed astronomer, who made an income through book sales and lecture tours. Valier had gained an audience, and his book on space travel sold fairly well. Many Germans were increasingly fascinated by a futuristic idea, which Valier promoted.

To Ley's surprise, Valier's book contained few original ideas. Instead, *The Advance into Space* simply popularized the theories of a professor named Hermann Oberth, the German "father" of rocketry. All credit went to Oberth's book, *Die Rakete zu den Planetenräumen (The Rocket into Interplanetary Space*, 1923). Ley had not heard of Professor Oberth, but he recounted what followed: "Intrigued—and with no inkling of how this would change my life—I soon saved enough to buy Oberth's own book. The clerk cannily showed me a similar book by Dr. Walter Hohmann—*The Attainability of Celestial Bodies*. I bought both at the penalty of being so broke I had to walk home, some three and a half miles." When Ley finally arrived home, he eagerly opened the books. He recalled, "I got a shock—both books were almost incomprehensible!" Despite his scientific background, "the principles developed in those books were just barely understandable." He exclaimed: "And the equations!" On Oberth's book, he commented, "As far as the general, even the interested, public was concerned, the book might just as well have been printed in Sumerian characters." Yet Ley persevered: "I studied the book over and over until it made sense."[12]

The core ideas of Oberth's self-published (and rejected) doctoral thesis excited Ley. Oberth attempted to prove four assertions: (1) "With the present state of science and technology it is possible to construct machines that can climb higher than the earth's atmosphere"; (2) with further development it would be possible to escape the gravity of earth; (3) these machines could safely transport a human into space; and (4) this technological accomplishment could both pay for itself and happen in the coming decades. Oberth also ruminated on the military potential of spaceflight technologies, such as a giant space mirror that could redirect and concentrate sunlight to destroy enemy cities. This nationalistic and militaristic sentiment would unite many of the early German rocket enthusiasts, although some simultaneously voiced internationalist rhetoric.

After Ley worked through the calculations, he realized that Oberth was the pioneer of an emerging field of rocketry. Germany stood at the edge of a new frontier of scientific exploration. Ley understood why Valier's book was gathering a small following, as there was clearly a need for a popular and intelligible book. Valier had attempted to translate incomprehensible jargon for a broader audience. Ley must have admired Valier's early success. Yet in other respects, Ley almost immediately disliked Valier's work and his public persona. Many of Oberth's concepts had been mangled or sloppily presented. Ley recalled: "Valier's book . . . was full of well-meant but ridiculous illustrations. . . . A great deal of the book was in fine print ('to be skipped by the lay reader'), full of calculations, most of them made by Oberth. In spite of all of these faults it sold at a fair and steady rate."[13]

The quality of Valier's popularization inspired Ley to compete with him. Ley explained: "With the enthusiasm peculiar to that age [of nineteen] I decided that I could do better than Valier." Not only would

Ley translate the complicated equations of Oberth, but he would also "simplify Valier's book in turn." This endeavor marked the beginning a writing career. It also documents how Ley investigated sources and repackaged dense material for popular consumption. This first endeavor highlighted a methodology of research that remained with him for the rest of his life. Without little formal training in a scientific or an academic field, Ley studied specialized texts, complex diagrams, mathematical equations, and primary historical sources. Often he discovered that beneath the complex mathematics, scientific language, or academic jargon lay a very simple and understandable idea.

Oberth had written a complex book, aimed at a scientific audience. In Ley's perspective, Oberth had done little to advance his cause in the public sphere. Ley later recalled: "No scientific fact or theme is too difficult to be explained to the intelligent outsider. If somebody says that this or that cannot be explained to the layman I understand this to mean that this person either does not have enough factual knowledge or else insufficient skill as an interpreter; often both." Oberth lacked the skill of communication. Valier lacked the scientific understanding. Therefore Ley would do what neither could do sufficiently. He would gut out the confusing diagrams and replace the equations with words. He would bring Oberth's vision of cosmic travel to the people.[14]

Journey into Space

Ley's quest to write a "small and formula-free" book produced *Die Fahrt ins Weltall* (*Journey into Space*, 1926), a sixty-eight-page treatise. In addition to explaining the theories of Oberth, Ley added his own perspectives on the possibility of alien lifeforms and the survivability of humans on other worlds. He then outlined the dangers of space-flight as well as the fundamentals of rockets and space travel. While relentlessly promoting the ideas of Oberth, the book culminated in a comparison of Oberth and his American counterpart, Dr. Robert H. Goddard, whose *A Method of Reaching Extreme Altitudes* (1919) had circulated in Germany. Goddard's often secretive experiments and brief publicity in newspapers were topics of endless speculation. Ley obtained his information about Goddard from Oberth's appendix, while associating Goddard with powder rockets. Ley argued that liquid fuels were clearly the next step forward. Oberth's ideas paved the way for a future of cosmic exploration. The direction of Goddard's research led nowhere, according to Ley.[15]

In hindsight it is easy to view Ley as a spaceflight advocate who grossly underestimated the difficulties, complexities, and costs of space travel. He expressed optimism regarding existing technologies and theories. Simultaneously, he minimized the dangers of velocity, meteorites, and cosmic radiation. The dangers and obstacles would be overcome. Engineers would discover solutions. The book concluded with a prophecy. When the first rockets escaped the earth's atmosphere, wrote Ley, "mankind, which physically and spiritually reigns upon the Earth, has taken a new step into a new age . . . THE AGE OF THE CONQUEST OF SPACE." It would be a transcendental epoch in human evolution.[16]

Ley's *Journey into Space* was not a popular success compared to Valier's book. But the book did sell nearly a thousand copies per year between 1926 and 1932. "To my surprise," Ley wrote, "many people, including Oberth, said that it was actually better than Valier's—at any event it did what Valier failed to do, it told the whole story understandably and in as few words as possible."[17] The light sales did not upset Ley. The book established his credibility among many disparate individuals on the scene. It also established Ley as a competent popularizer. Despite those successes, he still did not commit himself to becoming a writer: "Even after having done my version of Oberth's book, I did not think my future would be in such writing. I planned as before to become a paleontologist, or perhaps an astronomer. But I had no taste for engineering, and I only toyed with the idea of becoming a writer." He later admitted that even his desire to become a scientist felt rather directionless, although still connected to a central quest to become an explorer:

> To tell the truth: the border lines of those sciences interested me more than the actual material . . . and all through the astronomical lectures I wondered about Svante Arrhenius' theory of living spores traveling through space. If one could only go to other planets and

check on that theory. But propellers do not bite in a vacuum and gravityless [sic] substances violated half a dozen well-established laws of physics.[18]

"Do you see," he asked, "how the books by Oberth and Valier fitted in?"[19]

A Political Detour into Extremes

In 1925 Ley made a political decision. He joined the Nazi party, according to a 1933 membership questionnaire for the *Reichsverband deutscher Schriftsteller* (RDS, Reich Association of German Writers). After the Nazi seizure of power in 1933, the RDS attempted to "Germanize" institutions and media outlets, in order to fulfill the goals of the National Cultural Chamber Act. A massive purge of civil society and the private sector had begun. Consequently many writers declared their loyalty to the Third Reich. Ley admitted: "I belonged to the NSDAP [Nazi party] (Membership no. 778) since its beginning until the middle of 1928." In 2014 science fiction editor Wolfgang Both discovered this document, along with other evidence of Ley's self-identification as a protestant Aryan. Historian Michael Neufeld confirmed the identity of a Berlin-based member named Willy Ley, who joined in late July 1925, before being reported as no longer registered in late October 1928. Due to Ley's age, it is unrealistic to assume that his identification of the NSDAP's beginning relates to the true beginning of the Nazi party in 1919–1920. Rather, it refers to a reestablishment of the party under Adolf Hitler in 1925. There may have been lectures, nature hikes, or other semi-scientific activities that attracted Ley. It could be as simple as his bank supervisor's political leanings. Or Ley may have been an enthusiastic supporter. It would be difficult for him to ignore the anti-Semitism of party newspapers or Hitler's recently published *Mein Kampf*.[20]

Ley's motivations are unknown. Evidence suggests some degree of enthusiasm for the Nazi party. First, Ley began writing articles for the *Völkischer Beobachter*, the official organ of the movement. His articles appeared in issues with anti-Semitic, racist, or militaristic headlines.

However, Ley's contributions were apolitical. He wrote articles on technology, natural history, and mysterious legends. Second, science fiction editor John W. Campbell recalled that Ley attended a party rally in Munich, which accounted for Ley's ability to impersonate Hitler. This attendance cannot be corroborated. Third, Ley later made a statement to friend Robert A. Heinlein: in a 1940 letter Ley offered his opinion on the characters in the journal *Paris Gazette*. Regarding both Jewish and non-Jewish exiles in Paris, Ley wrote that "in general, I dislike those people." He identified their "lack of ability to assimilate themselves." Ley then sympathized with their plight of simply waiting to return home, yet he commented, "I still don't like those types." Again he referred to all German exiles, not simply German Jews. Nonetheless, he added this cringe-worthy joke about young German Jews who usually faded from the intellectual scene "after one possibly brilliant but at any event doubtful meteoric performance in the realm of letters." Other comments poke fun at "the arche-type [*sic*] of an intellectual Bavarian, a type that manages to be just important enough to be tolerated in spite of its [*sic*] bad and offensive habits."[21]

Otherwise it is difficult to explain his political leanings. As the remainder of this chapter illustrates, Ley's Nazi party membership coincided with his increasing involvement in circles of struggling engineers and bitter veterans. They voiced support for a unified scientific community, in which all German-speaking scientists and technicians demonstrated the superiority of the German mind. Ley participated in that nationalistic effort to bring German scientists together, in spite of arbitrary borders. At the same time, many of the rocketeers would consume right-wing science fiction, which provided fantasies of revenge. These camps of engineers and writers influenced Ley. Through his growing association with militarists, nationalists, and unemployed engineers, he may have sympathized with their political leanings before becoming far more internationalistic in 1928. He would transition from a cause of uniting German minds to fostering international cooperation among all types of visionaries. Eventually the cause of spaceflight transcended borders.

The Museum of Mars

In fall 1926, as Ley reviewed corrections to *Journey*, a friend told him of a Mars exhibition at the observatory of Treptow in Berlin. Ley accompanied friends to view telescopic photographs of the red planet as well as browse the literature for sale. Ley noticed how similar were the museum of Mars on earth and the "Museum of Earth" on Mars, as depicted in Lasswitz's *On Two Planets*. Ley probably hoped to see photographs of the allegedly engineered canals on Mars. He must have been disappointed by the blurry and non-definitive exhibits as well as the lack of readable, popular books. Nonetheless, the museum enchanted him. He longed to unravel the mysteries of Mars.

Soon after the event Ley wrote a supplement to *Journey into Space*, called *Mars, der Kriegsplanet* (*Mars, Planet of War*, 1927). As a small paperback treatise, *Mars* would provide an "understandable description of the results to date of astronomical research about the neighboring planet." Not only did Ley recount the history of astronomical discoveries and theories, but he also outlined the recent controversies over canals. In this regard the book served as a sequel to *Journey*. "In the first part [of *Journey into Space*]," Ley wrote, "I showed that it would be possible for people to live on an alien world for a brief or extended period. Then, in parts II and III, I presented the technical details as to how one could enable people . . . to reach such strange planets." With this new sequel, Ley "strove to make contemporary knowledge about Mars easily understandable, readable, and accessible for the broadest strata of the population." He hoped that the book would provoke the imagination of his readers. He would serve as their knowledgeable tour guide.[22]

As Ley took readers on a trip through the history of astronomy, he praised human curiosity and the bold thinkers of the past while he examined historical manuscripts and books. Astronomy and the quest for other worlds sparked the boldest of changes in Western thought. In this narrative, seventeenth-century astronomer Johannes Kepler was an early pioneer. Kepler combined rigorous reason with imaginative leaps. He also dreamed of spacecraft and a trip to the moon. Galileo and other bold adventurers did likewise, aiding the destruction of traditional worldviews. The book then moved quickly into the nineteenth century, Ley's comfort zone. Ley had grown up reading the works of nineteenth-century popularizers, such as Cammille Flammarion and Richard Proctor. Their earlier accounts influenced Ley's presentation of the material, albeit with less refined tactics. Much of the text is simply descriptive, as Ley creates a "who's who?" account, along with a list of steps in the right direction. Nevertheless, the book also includes sections relevant to the cultural history of astronomy, particularly on the role of media, hoaxes, and science fiction. Ley leaves room for unsolved mysteries. On other points, scientists had discovered the truth. The darkening areas of Mars indicated vegetation, according to many astronomers.

The book concluded with a discussion of "Ignorabimus," a Latin term referring to the limits of human knowledge. Ley asked, "Will we ever know? Does Mars harbor life?" He expressed doubts about earthbound astronomy. He also doubted the possibility of communicating with Mars through existing technologies. If the mysteries of Mars were to be solved, then it required manned space travel. Ley argued that the "rocket ship will unveil [the mysteries] for all of us, achieving what has always been dreamed, yet considered impossible: to carry men to Mars." This trip would discredit "the mocking laughter of certain cultured and educated men." Ley then expressed his high hopes for a club that had recently formed in Vienna. Experiments would soon begin. If he could infiltrate the inner circle of experimenters, perhaps one day he might be among the next generation of explorers.[23]

The Society for Space Travel

In spring 1927 Ley received a letter from Valier, who suggested that a club be formed to finance the rocket experiments of Oberth. Ley soon met Valier for first time during a Berlin lecture tour. Ley described his first impressions of Valier as a kindred soul. By this point Ley was aware of Valier's more eccentric writings on occult sciences. He may have known about Valier's goal of using rocketry to confirm Hans Hörbiger's "World Ice Doctrine," which posits ice as the fundamental substance of cosmic processes. They had a cordial breakfast together.

Valier mentioned that a Munich professor had lectured about English war rockets, developed in the nineteenth century. Ley remembered, "Valier suggested that I might check whether rockets had a use in history other than as mere fireworks . . . [which was] a fact scarcely suspected before." This challenge intrigued Ley. He began to research the history of rockets, as Valier continued his publicity tours.[24]

A space travel club materialized. The founding meeting of the Verein für Raumschiffahrt (VfR, Society for Space Travel, also called the German Rocket Society) was held in July 1927. As noted by historian Frank Winter, the VfR eventually became "the most prestigious of all the space travel organizations." Ley recalled, "Our sights were set high from the very outset." The group's ultimate goal was ambitious: "The purpose of the union will be that out of small projects, large spacecraft can be developed which themselves can be ultimately developed by their pilots and sent to the stars." In other accounts Ley described more ideological goals. Their main purpose included "spreading the thought that the planets were within reach of humanity if humanity was only willing to struggle a bit for that goal." Ley also commented in 1940: "The Society had been founded in 1927, mainly as a kind of scientific debating club." It is interesting that early accounts of the VfR stress the informal nature of a rocketry fan club, while later accounts stress the scientific and technological goals of the organization.[25]

At least nine rocket enthusiasts, including Valier, met in the back room of a Breslau tavern. Machinist Johannes Winkler accepted the presidency and his role as an editor of *Die Rakete*, the club's newsletter and journal. Ley was neither present nor listed in the charter. Winkler and Valier agreed to assemble a list of names, in order to send out club invitations. The membership incentive was its journal, *Die Rakete*, an interesting mixture of reports, technical articles, and serialized novels. In some ways it could be read as a scientific journal, offering detailed accounts of developments in the field. At other times, it either covered sensational rocket stunts or provided a new installment of a science fiction novel. It also printed poetry and jokes. VfR founders hoped that subscription dues would finance experimentation. Thus they had an incentive both to entertain and to educate subscribers.[26]

Initially, the VfR was successful. "The growth of the VfR was rapid,"

Ley stated, "Within a year it acquired almost five hundred members, among them everybody who had ever written about the problem in Germany and in neighboring countries." Despite hopes for experiments, the group's initial contributions to the cause of spaceflight lay in the realm of media and publicity, not in sites of research and experimentation. The VfR remained committed to publicity throughout its existence.[27]

For many historians Ley is a central figure in the creation and growth of this international (although mostly German-speaking) scientific community. It is true that Ley became instrumental in the group's correspondence and publicity. He became one of "the" spokespersons for the organization as well as a co-editor of the newsletter. Yet Ley's role in the early days of the VfR has been inflated over time. He was absent from the pages of *Die Rakete* prior to 1928. He later insisted that he was a co-founder of the organization, due to his work as an international correspondent. There is little evidence to support this claim. We can only speculate as to why Ley later inflated his early connections to the organization. In 1927 Ley may have viewed the centrality of Valier with a skeptical eye. One thing seems clear. When Valier and others founded the VfR in 1927, Ley was attempting to reach a wider audience as a science writer on natural history. He was branching out.

Rockets of the future were interesting. But the creatures of the past were marvelous.

The Book of Dragons

After writing an appendix for a book on the Ice Age, Ley spent much of 1927 researching and writing *Das Drachenbuch* (*The Book of Dragons*, 1927). In 208 fact-filled pages Ley explored the natural world of reptiles, amphibians, dinosaurs, and other amazing but misunderstood creatures. Chapters included "The Struggle with the Dragon," "The Great Sea Serpent," and "The Survivors of the Australian Bush." Like the earlier books of Dr. Cell, *The Book of Dragons* educated and entertained readers, while often debunking myths and unpacking legends. Ley explained the goal of the book: "These animals are seen as either disgusting or toxic" by most Berliners. If people would simply see the creatures up close in the many wonderful aquariums of the city, they would suddenly view them differently. In the illustrated pages that followed, Ley served as a zoological tour guide, walking readers through Berlin's local museums. Other chapters, such as "The Great Sea Serpent," encouraged readers to wonder about future discoveries of stranger creatures and other living fossils. "Well, everything is possible," Ley concluded. Readers are encouraged to experience a sense of wonder, mystery, and awe. The beauty and the terror of nature was on display.[28]

The Book of Dragons reflected Ley's evolving writing style, as he mixed scientific explanations with interesting stories about the people who discovered creatures and fossils. He imitated the styles of his favorite science writers. He also experimented with humor, which was absent in his earlier books. His language became playful, especially when referring to dragons. Often terms or characters from fairy tales are inserted into the narrative, while illustrations could be equally playful. Occasionally Ley offered reflections, as if speaking to the reader in a more personal tone. Although his writing style would evolve in later decades, every key element was present in his first book-length work. *The Book of Dragons* educated and entertained. It debunked myths and combated popular misconceptions. It excited readers about the wonders of the world. It placed its audience in exotic locations, including the distant past. The audience encountered fascinating landscapes that seemed alien and otherworldly. As imaginative explorers, these readers discovered playgrounds for wonder and awe. With Ley as their guide, they became fellow explorers. On a grand adventure, anything was possible.

The Possibility of Space Travel

After Ley spent 1927 branching out, he returned to the subject of space travel in 1928, when he edited a collaborative book on rockets and spaceflight. This was the period when he became more involved with the VfR. "My plan was about as follows," he wrote: "First, get all the people who had contributed ideas together and make them write a book in collaboration." This endeavor produced *Die Möglichkeit der*

Weltraumfahrt: allgemeinverständliche Beiträge zum Raumschiffahrtsproblem (*The Possibility of Space Travel: Understandable Contributions on the Problem of Interplanetary Travel*, 1928). It included chapters written by most of the notable German-speaking individuals in the field, including Oberth. Additionally, Ley later claimed to have reached out internationally, particularly to Robert H. Goddard in the United States. According to Ley, Goddard did not reply.[29]

Possibility had obvious goals. Ley sought to produce "a readable book which would convince a great number of people, not precisely the man in the street, maybe, but engineers, teachers, the higher-ups in the civil service, and so on." If scientists, engineers, and theorists wrote in clear and intelligible language, then they would excite fellow Germans to support the cause. To accomplish this task, Ley arranged the material in a dramatic way, moving from his own speculation about extraterrestrial life to contributions on the history of science fiction, the future of space stations, and the underlying principles of space travel. Ley ended the book with a plea to the audience to believe in spaceflight. He wrote: "It is no longer fantasy . . . to speak of a ship. . . . Space rockets are the future!"[30]

Other justifications for the book had a nationalistic tone. Ley argued, "My present hope is that . . . from this German rocket book a German space ship will emerge." Tellingly, he did not distinguish between his German, Austrian, and German-Romanian contributors. They were all Germans, regardless of government. This perspective acknowledged a Greater Germany, which existed beyond arbitrary borders. Ley also implied that science served as a unifying cultural force in German-speaking areas. Nationalism, patriotism, and pride could lead to the conquest of space. Germany would prevail due to the brilliance of German theorists and engineers.[31]

Ley had these nationalistic and lofty goals in mind as he sought to excite the public. In his view he helped to make rocketry respectable as an emerging field of engineering. Indeed, the book was successful. Ley reminisced in 1943: "It even sold well in spite of the high price, the equivalent of five dollars." However, Ley believed that his efforts were sabotaged at the moment of publication. Instead of a modest but respectable degree of public support, a wave of "colossal nonsense" distracted the public. The man responsible for "the greatest possible misunderstanding and stupidity" was none other than Max Valier.[32]

A Popular Fad and a Founder's Betrayal

The years 1928 and 1929 marked the peak of a rocketry fad in Weimar Germany. Historians have offered several explanations for the flourishing of rocket-related media and spectacles. Frank Winter identified philosophical roots in "Lebensphilosophie" (philosophy of life), a rather obscure and anti-rationalist camp of romantic philosophy. According to historian Michael Neufeld, the fad is better explained by a combination of nationalism, widespread beliefs about technological progress, and a vibrant consumer culture. Tom Crouch put the fad in more generalized terms: "politics, economics, and culture had paved the way for the coming of the rocket." Despite the interpretative differences, most historians agree that nationalism played a key role. Through a collective celebration of German technological might, many Germans gathered in crowds to witness new technologies on display. Often the events explicitly or implicitly celebrated the future resurgence of Germany as a scientific and technological power house. For a short time rockets had their place in the limelight. From the words of reporters and science fiction writers to the images in German advertisements, the rocket and the rocket plane signaled the dawn of a new era. Germany would lead the world into the next frontier. Writers expressed a triumphant nationalism. Despite the humiliations of the past and present, the German mind would prevail.[33]

Historians can easily point to these larger trends. However, it has been more difficult to determine the influence of specific individuals, such as Ley or Valier. In Ley's perspective the VfR worked to make Germany "the most rocket-minded country." The engineers and popularizers succeeded in exciting Germans to support the cause. From books and articles to lectures and public exhibitions, the advocates campaigned. The VfR grew. Yet, as Neufeld argues, none of these efforts made a deep impression on the public until Valier made headlines with rocket stunts. Neufeld's point is well taken, although historians lack a broader cultural history of the fad that explores consumer culture, along with a vast array of media that flourished. Perhaps the rocketry fad can be attributed to a flourishing of romantic, popular science during the stabilization period.[34]

Undoubtedly Valier's stunts grabbed headlines, after he persuaded automobile tycoon Fritz von Opel to construct and race "rocket cars." These vehicles utilized powder rockets attached to the rear. On May 23, 1928, crowds of Germans cheered as Opel's "Rak II" soared down the Avus speedway at 125 miles per hour. Prior to the race an engineer announced: "Within a few years it will be possible for Berliners to travel across the ocean within five hours—to breakfast at home, lunch in New York, and to return in time for the opera." Opel also announced over radio: "In the end, we may try to penetrate into space, but that is still a dream." Nearly every German newspaper reported on the events, while connecting Opel's success to innovations in rocketry. Reporters lauded Opel's designs for rocket planes as the next step. Eventually Valier's powder rocket stunts extended to other types of vehicles, including planes, sleds, and bicycles.[35]

Ley participated in these events as a skeptical onlooker and hesitant journalist. He despised the "big, carefully staged show." He viewed the rocket car stunts as dangerous and detrimental to the cause of the VfR. The chief sin of Valier and Opel involved the use of powder fuels. The fundamental contribution of Hermann Oberth and the central quest of rocket research related to the development of liquid fuels. A powder rocket was the "plaything of a pyrotechnist" [sic]. Theorists had shown that "powder rockets . . . were almost the most inefficient type conceivable." Ley recalled: "We had gone to extreme lengths to explain the numerous advantages of liquid-fuel rockets to anybody who would listen—and Valier went and made publicity for von Opel with commercial powder rockets!" Ley also believed that the publicity damaged the credibility of the VfR, when Valier publicly associated the organization with the stunts. Ley recalled: "Small wonder that this victory did not make us very happy. . . . The efficiency of these runs had been below 1 per cent! The expense had been fantastic."[36]

Ley understood Opel's role in the affair, because Opel "saw an opportunity for purchasing unlimited publicity." Yet Max Valier, as the most famous member of the VfR, participated in "headline stunts"

an international republic of letters. Ley stopped contributing scientific articles for the Nazi press. Instead, his short articles began to appear in leftist newspapers such as *Vörwarts*, the official organ of the Social Democratic Party. Ley also began to correspond with foreign enthusiasts. His attitude changed as he conceived of rocketry in international dimensions. Ley became a key intermediary in a larger international movement.[41]

Ley's internationalism might be related to the classic perspective of historian Paul Forman, who argued that scientific internationalism, as it was voiced in Weimar Germany, became a convenient smoke screen for nationalistic science. While scientists paid lip service to the ideology, they subverted its classical premises. Forman concluded that genuine internationalism did not exist in the Weimar scene. Ley's subsequent correspondence with interplanetary societies and foreign enthusiasts might be characterized as "a supranational agreement on the ground rules." The international nature of the VfR bolstered the prestige of German scientists and engineers, who encouraged a united front of research. In reality the VfR was part of a movement that thrived on nationalism and science fiction fantasies of revenge. However, Ley's internationalism was genuine. He despised the secrecy of Goddard (and later Oberth). He attempted to disseminate all knowledge throughout the German-speaking world and beyond. He even wrote equations on international postcards. His internationalism was not a smoke screen. Ley grew eager for international cooperation in the conquest of the next great frontier.[42]

There is a direct correlation between Ley's internationalism and his greater involvement with the VfR. Ley recalled, "I did more and more of the club's work," which included secretarial and editorial duties. Ley also claimed that by the summer of 1928 he "joined Winkler in the editorship of the *Rocket*." Although the journal does not indicate co-editorship, it seems clear that Ley became the group's chief international correspondent, as he built a transnational network of theorists, engineers, science fiction writers, and rocket enthusiasts. One might argue that Ley subverted the agenda from nationalism to internationalism. After Winkler stepped down as president, Ley became vice president (while Oberth became a distant and disinterested president). In a later interview, Ley joked, "I was vice president, and for a long time, there was no president." At this point, most of the organizational, secretarial, financial, and publicity responsibilities fell to Ley. He also took on many duties associated with the publication of the *Rocket*. The publication soon folded due to various reasons. According to his later accounts, the VfR continued to grow.[43]

In 1929 Ley also revised his *Journey into Space*. The book offered a new foreword, written by Oberth, who claimed that Ley's 1926 edition was the "first truly popular German rocket book." Ley returned the praise in kind. Yet the second edition was more than a simple popularization of the theories of Oberth. Ley offered a completely reworked edition with thirty illustrations. Additionally, the book contained new material on the history of rockets. It attempted to put the rocket into a broader historical context. In fact, the history of the rocket was "a story all its own." The text also outlined the accomplishments of Oberth's predecessors. The history of rocketry had a long list of pioneers and founding fathers. Ley wrote, "Since 1907, Hermann Oberth busied himself with a large problem, just as Walter Hohmann had. In St. Petersburg there was even a lively debate between Esnault-Pelterie and Ziolkovsky." Although this debate did not actually occur, Ley's reporting indicates a broadening international focus. In his perspective the history of rocketry included an international effort that the Great War "tore to shreds." Ley sought to rebuild the networks of communication and exchange. He openly shared knowledge.[44]

The book also indicated Ley's evaluation of the potential of military rockets. Time and time again, the war rocket emerged as an ineffective novelty, used to supplement more effective artillery. As a stand-alone technology, it was practically worthless as a short-range device that offered very little payload. At best it could be a useful signaling device. History proved that powder war rockets belonged to the past. The liquid-fueled space rocket belonged to the future.

The Woman in the Moon

October 1929 marked the zenith of Germany's rocketry fad. The pinnacle of Ley's success surrounded an event "which seemed to have little to do with science but which was to have lasting influence." "That event," Ley recalled, "was the premiere of a film, on October 15, 1929." It was titled *Frau im Mond* (*Woman in the Moon*). Fritz Lang, the film's director, had already become famous for *Metropolis* and other films. In 1928 he was inspired by both the sensational publicity and serious books. So many Germans were excited about a future of interplanetary travel. A film that depicted the German conquest of space could be profitable and awe-inspiring. It could also be visionary and scientific.[45]

News of Lang's film was a godsend for the spaceflight advocates. According to Ley, it was "almost impossible to convey what magic that name had in Germany at that time." Ley described the typical premiere of a Fritz Lang film in Berlin:

> The first showing . . . was something for which there was no equivalent anywhere as a social event. The audience . . . comprised literally everybody of importance in the realm of arts and letters, with a heavy sprinkling of high government officials. It is not an exaggeration to say that a sudden collapse of the theater building during a Fritz Lang premiere would have deprived Germany of much of its intellectual leadership at one blow, leaving mostly those who for one reason or another had been unable to attend.[46]

Lang's film would be a monumental boost to the cause, according to Ley. It "meant a means of spreading the idea which could hardly be surpassed in mass appeal and effectiveness." Ley and others also hoped that it might generate sizeable funds for rocket experimentation. Even more encouraging to Ley was Lang's choice of Oberth as a scientific consultant. Clearly, here was a chance to hit it big, well beyond publicity stunts with solid-fuel rockets. The movie would credit Oberth as the real genius behind the rocket ships of the future. Surely that fame would translate into real funds for experimentation, Ley believed.

In his later books Ley told and retold a series of humorous events surrounding the film and Oberth. Ley recalled, "At first they had Max Valier in mind as a scientific advisor." Ley continued, "For some reason, he [Lang] decided that Max Valier was not the man for him, and he needed somebody better, so he wrote a long letter to professor Oberth. . . . And then Oberth came to Berlin." Being accustomed to small Transylvanian towns, "unhurried small-town intellectuals," and leisurely study in Heidelberg, Oberth now "found himself in the very spot where the apparent turmoil of a big city appears wildest." "Suddenly plunged into the strange atmosphere of fast-moving, efficient, flippant, and sophisticated Berlin," Oberth became confused by his surroundings. "His mental make-up," Ley wrote, "was strange indeed." First came a period of "astonished disbelief that everything can be so different." He arrived in foul spirits, stubbornly refusing to compromise any degree of scientific accuracy. Oberth distrusted everyone and everything around him. He also voiced his disapproval of Berliners, who "had no soul and were German-speaking Americans, hunting money all the time." According to Ley, a "mystic inclination . . . naturally transformed Oberth into a Nazi in due course."[47]

Ley intervened, asserting himself not only as Oberth's guide in Berlin but also as a friendly intermediary, who could help Oberth understand the local dialect. Ley tried to convince Oberth not to argue. "I began to needle Oberth," he claimed. "Professor," he said, "these are the movies, not just the movies even, Lang himself. Money doesn't matter here, this is where you can get the cash to transform your formulas into reality." While Oberth "could not understand people," he would get angry at Ley for trying to advise him on manners and social skills. Oberth continued to struggle with the "ultrarapid" dialect, a complicated transportation system, and the foreign customs of the movie industry.[48]

Despite Oberth's difficulties, he was able to secure some funding from both the Ufa Film Company and Fritz Lang himself. The contract specified that Oberth construct a small liquid-fuel rocket, to be launched during the film's premiere. Although Ley claimed that "it was not in itself a bad scheme," a series of haphazard blunders ensued. First, Oberth hired a "Hitler-voiced unemployed engineer" named Rudolf Nebel as first assistant. Nebel would soon dominate the scene as an important figure. Ley would later refer to Nebel as a deceitful con man who lacked experience as an engineer. Instead, he was a "salesman of mechanical kitchen gadgets . . . not the man Oberth needed." Next, Oberth hired an exiled Russian student named Alexander Shershevsky, who was both a "frenetic communist" and "lazy by nature."

According to Ley, Shershevsky "would much rather have discussed the concept of infinity in mathematics, the importance of radicalism in politics, and the great work they would be able to do if they only were at the Central Aero-Hydrodynamical Institute in Moscow."[49]

Ley continued: "This trio, consisting of a bewildered theorist, a professed militarist, and a Bolshevist accidentally in disgrace, worked together, or tried to." Ultimately they failed. Oberth made some experimental progress on his "Kegeldüse" design, before an explosion nearly cost him his eyesight in one eye. Other events caused delays. Running out of time, Oberth "rapidly approached a nervous breakdown." He fled home for a week and then briefly returned to make legal threats to the Ufa film company. Then he left for good. According to Ley, Oberth later told the president of the VfR "that he had not been accountable for his actions," because the explosion "had given him all the symptoms of shell shock." The entire engineering adventure resulted in failure. Ley later admitted, "Oberth, I regret to say, was not the proper man to do it. As a matter of fact, such a man did not exist at all. There was nobody at that time who had sufficient experience with liquid fuel rockets." Oberth, in particular, "had no idea of how to go about it." After all, "he was a theorist, not an engineer." Nevertheless, the film depicted his vision of a future of human spaceflight.[50]

What is amazing about this story is how very little Ley wrote of his own role in the publicity and content of *Frau im Mond*. Instead, he always credited Oberth as the scientific mind behind the film's realistic depiction of a rocket flight to the moon. He recalled: "The spaceship shown was not some 'artist's conception' but a design by Oberth. He had calculated all the dimensions, and the model shown in the movie was a precise scale model." The film also credits Oberth as the sole technical consultant, while Oberth later recalled that Ley "was only an author." To this day most historical accounts of the film credit Oberth exclusively.[51]

Consequently it is interesting to read Fritz Lang's memories of the events. In an obituary of Ley, Lang attempted to set the record straight: "I met him [Ley] in 1927. He was 21 years old and had already written two books on space travel which had been published in Germany in 1926. . . . I contacted him because I planned to make a picture 'Frau im

Mond.'" If this recollection is accurate, then Ley and Lang first met in the fall of 1927. Ley's two books could refer to *Journey* and *Mars*. Lang also recalled, "I was very much impressed with Willy Ley from the beginning on as much with his humility as with his tremendous technical knowledge of the subject." Lang's remark about Ley's humility may have a deeper meaning in relation to the film. Lang also recalled that they "became friends in the truest sense of the word." This friendship would last a lifetime. Lang continued: "He suggested we call in Prof. Hermann Oberth [who] came to Berlin, and with the great help of Willy Ley a concept of space flight was developed which I portrayed in my picture." Lang then credits Ley as the film's consultant:

> The work he [Ley] had done as consultant and adviser to the film "Woman in the Moon" was amazing. The models of the spaceship, really a highly advanced model of a rocket, the trajectories and the orbits of the modular capsule from the earth, around the earth and to the moon and back, were so accurate that the Gestapo confiscated not only all models of the spaceship but also all foreign prints of the picture.[52]

Lang credits Ley as the inspiration for the film, saying, "Willy Ley had already originated the concept of space flight of a rocket from the earth to the moon, which enabled me to make my film."[53]

It would be easy to dismiss Lang's account as an embellished obituary of a dear friend. The language was exaggerated. Much credit still goes to Oberth, because Ley was trying to popularize Oberth's ideas and concepts. Nevertheless, the tale is quite plausible, because it reveals a recurrent feature of Ley's entire career: he could become the man behind the curtain, working with movie directors, television executives, and other cultural producers. As he simultaneously shaped the content of the production, he credited the spaceflight imagery to Oberth (or later von Braun). With *Frau im Mond*, there are reasons to suspect that he was deeply involved with the planning of the movie. If there was a book that inspired Lang or his wife, the scriptwriter, it was neither Oberth's unreadable mathematical treatise nor his more readable book, published later in 1929 as *Wege zur Raumschiffahrt* (*Path to Space Travel*). Instead, it was Ley's or Valier's translations of Oberth's concepts, either in person or in book form. It is possible that Ley heard about Valier's involvement before he intervened to sabotage the deal. Regardless of the details, Lang consulted Ley, who suggested that they invite Oberth. Most likely, Ley and Lang wanted to attach Oberth's name to the film to give it some degree of scientific respectability.[54]

After Oberth arrived in a foul mood, Ley accompanied him to the studio to tour the set. While Oberth proved difficult to work with, Ley and Lang got along splendidly. Ley later claimed, "I got involved in the film myself" after the arrival of Oberth. Ley's version of these events may have hidden his contributions to the film in order to give Oberth full credit. Whereas his early accounts of these events entirely credit Oberth, a later account in the 1960s displayed less humility. For example, Ley admitted: "I wrote most of the scientific publicity for the film." It seems fairly certain that Ley was an active part of an "unparalleled advertising campaign." While he wrote articles for the popular press, the Ufa film company sold postcards, posters, and even "rocketlike kaleidoscopes through which peephole you could see 'the woman in the Moon,' with bare arms reaching for the stars." Ley served as the film's secret publicist.[55]

Ley had legitimate reasons to feel proud of what he had accomplished, particularly on the night of the film's premiere. He attended the event at the invitation of Lang. Ley recalled the scene as stunning, despite the absence of a publicity rocket launch. Ley added, "From Hugenberg to Einstein, you could, paradoxically speaking, see no one who wasn't there." Ley then described the most important scene of the film. As a multi-stage rocket is revealed to a contemporary-looking crowd, the spectators cheer wildly. After a dramatic countdown, the rocket launches and the bold adventurers struggle to cope with the physiological effects of space travel. According to Ley, the audience found the scene spellbinding:

> There is without question no other scene, either on Earth or on the Moon, that would have ruffled the poise of this cool, reserved, expert audience—these journalists, scholars, diplomats, men of affluence, and film stars. In the face of these outstanding technical achievements, the audience exploded. Electrified, carried away. The fiery

jets of this film rocket swept away their carefully prepared skepticism, indifference, and satiety with the same speed with which the rocket raced across the screen.[56]

Frau im Mond gave them "a small glimpse of the tremendous possibilities."[57]

Ley was pleased that a silent film could do so much to stir emotions and excite a crowd to envision a future of humans in space. It would only take time for technology to catch up to the power of media. Fantasy would inspire reality.

Tactics and Tensions

Ley's formative years were characterized by an increasing commitment to popularization, the coordination of experts, and the mixing of genres. Ley's commitments and tactics during Germany's rocketry fad illustrate much about his underlying belief: science was open to both a larger international community of experts and a Western audience who could finance experiments. With the right intermediaries and translators, the specialists and the public could come together. The public could even become the specialists' patron. Imagination became the most important tool for converting audiences and funding experimentation. Oberth and other theorists would achieve little by speaking only to one another. The specialists would be lost in isolation. They needed an effective coordinator and popularizer. Their cause needed a publicist. Ley had stepped into the scene. He shaped his public persona accordingly.[58]

In Ley's view there was a right and a wrong way to inspire the public. A futuristic fantasy like *Woman in the Moon* aided the scientific and engineering crusade, despite depicting an oxygen-filled atmosphere on the moon. A movie could take a few artistic liberties. Oberth's refusal to compromise any degree of scientific accuracy was unreasonable to Ley. Oberth needed the public, and the public needed to be excited. Some latitude with popular media and sensationalism was necessary and even desirable. Yet when it came to Valier and Opel's stunts, Ley was far less forgiving. Although the rocket vehicles excited the German public, they did nothing to advance the cause, due to their reliance on powder fuels. Additionally, Valier increasingly spoke and looked like a charlatan and pseudoscientist. The public and the cause needed a more moderate spokesperson. Ley adopted a public persona that contrasted with Valier's more flamboyant style. In Ley's view, a popularizer could be a showman, if the show was respectable and generally honest. Most notably, the show could have lasting effects on public perceptions and popular support. Enthusiasm would lead to funding, which would undoubtedly lead to technological innovation.

Ley belonged to a camp of enthusiasts who had naïve views of technological innovation. Historians have rightly critiqued their optimism and expectations. Nevertheless, there is an important distinction between Ley and other advocates. As an amateur historian of science, Ley began to reflect on the influence of culture and how social hopes, intellectual dreams, and a broader context shaped science and technology. By no means were his views as well developed as the views of later historians. Yet he thought in similar terms. In his perspective the pseudoscience of the "Dark Ages" reflected cultural obstacles, superstitious beliefs, and a lack of imagination. The science of the Enlightenment also reflected a broader context of social beliefs and expectations. Culture affected both the development and application of science. For the field of rocketry, two factors must be maintained. First, the general public had to be excited, thereby creating a fertile environment for the dreamers and engineers. Second, the scene had to be international, open, and cooperative. A republic of letters among scientists could communicate through the universal language of science. Specialists would combine energies to overcome parochialism and borders. Together they would explore a new frontier after standing on the shoulders of giants.[59]

Paradoxically, Ley's internationalism had to confront the first moment of public excitement: Valier's public stunts and heightened nationalism accompanying the scene. Ley did not criticize the popular nationalism, but it is easy to imagine him feeling uncomfortable with the emerging tensions between his growing commitment to internationalism and the swell of nationalism during the rocketry fad. Perhaps Ley embodied the tensions of a broader scene. For example, the emergence of consumer and cosmopolitan culture could upset nationalistic sensibilities. Consumer culture was becoming international or Americanized, according to many cultural critics. Additionally, the glorification of technology could include a mixture of nationalistic dreams of revenge and internationalist fantasies of peace, diplomacy, and a "winged gospel." Scholars have documented the contradictory representations of the airplane. Few of these tensions have been explored in relation to rockets.

Although an analysis of Ley's internationalism does not reveal how widespread these tensions were among other enthusiasts and amateurs, it provides an interesting case study that begins to explore conflicting ideologies and contradictory practices. As we can see in the next chapter, these tensions exploded with the triumph of technologically minded nationalists. Ley became a key participant in a cultural and technological conflict. Prior to and after the Nazi seizure of power, Ley's scientific internationalism and organizational hopes would be pushed to a breaking point. He would perceive the death of a science. He would also fear for his life.

Death of a Science in Germany

Germany's rocketry fad peaked in the fall of 1929. A period of experimentation followed, in spite of the economic impact of the Great Depression. The field of rocketry progressed. Ley had many reasons to be hopeful. However, by 1934 the movement was dead, in Ley's opinion. The rise of the Nazis and the militarization of rocketry "killed" his scientific cause. How did Ley perceive these events? How did he view the relationship between science and politics? How did his own political views influence his activities and perceptions? These questions are very difficult to answer, because almost all evidence comes from Ley's later memoirs, written after he had left Nazi Germany in 1935. In these personal recollections his anti-fascism is clear. Yet in pre-1935 documents the evidence is more ambiguous, especially when Ley suspected that authorities were monitoring his correspondence. He often ended a newsletter and correspondence with the words "Heil Hitler!" Like other journalists and citizens, he reapplied for Nazi party membership.

Although historians must rely on Ley's later memoirs to reconstruct the events of this period, a few aspects can be verified. In spite of the Nazi campaign against international scientists, Ley continued to cultivate ties with foreign enthusiasts. He openly shared technical information. He looked for opportunities for foreign funding. He exchanged information with a Soviet group called GIRD. Ley also published mathematic equations and formulae in the American publication *Astronautics*. With the decline of the Verein für Raumschiffahrt (VfR, Society for Space Travel) he sought to consolidate all groups into a unified and transatlantic organization. His efforts as an international publicist were genuine. Both his correspondence and his first science fiction novel reflect his internationalism.[1]

By combining this evidence with later accounts, we can chart his commitment to scientific internationalism as well as the growing dangers for such a commitment. In Ley's perspective the cause of spaceflight required international cooperation, publicity, and the open sharing of information. After 1933 the totalitarian state demanded secrecy, paranoia, state control, and the persecution of scientists and engineers. When rocketry became militarized, everything fell apart, in his view. Totalitarianism, irrational politics, and pseudoscientific nonsense poisoned Germany's scientific and engineering well. For Ley the situation became desperate. Ley's perceptions of events did not always match the historical realities. Yet they established a worldview that greatly influenced his activities during the next two decades.

The Starfield Company

At the height of Germany's rocketry fad, Ley wrote an idealistic science fiction novel called *Die Starfield Company*. The book reflected his mindset in 1929, when he had such high hopes for the field of rocketry and the international scene. The novel is also remarkable for its contrasts with other German science fiction stories. It is important to recognize the diversity of Weimar science fiction. German fantasies of the future could range from left to right on the political spectrum. Yet many scholars still point to common themes that distinguished most German science fiction as mystical, nationalistic, and even proto-fascist literature, which combined fantasies of revenge with the glorifications of wonder weapons.[2]

From the technological fantasies of Hans Dominik to the lesser-known works of aspiring authors, German science fiction depicted the renewal of a Teutonic empire in the sky and beyond, along with the crushing defeat of Great Britain and France. These political fantasies could be incredibly nationalistic. For example, consider a passage from Otto Willi Gail's *Der Schuss ins All* (*The Shot into Infinity*, 1925), in which the main engineer shouts:

> The Dirigible, the Graf Zeppelin, years ago spread over the whole earth the fame of German spirit, German technique, and German work, so that our former enemies recognized that this nation was alive, despite all suppression. . . . And now the lofty music of German ability shall resound in the canopy of stars—to distant unknown worlds, the German colors shall shine and announce that *this nation lives!*[3]

Other novels openly advocated for war, as done by a character in Ludwig Anton's *Interplanetary Bridges* (1922):

> Once we have sunk one or two English or Japanese men-of-war [with our airship] . . . once we have shown the world that we have sharp claws and know how to use them, then, and only then . . . they will recognize us as an equal power, make commercial and political treaties with us and invite us to join their League of Nations. And then Germany's time will have come to maintain her old-time prestige against all the nations of the world.[4]

Although there are several exceptions, much of this literature combined machine dreams, nationalistic passions, and revenge fantasies. Quite often the genre distinguished itself by expressing antidemocratic critiques of bureaucracy, while simultaneously praising a dictatorial great leader, along with the harmonious *Gemeinschaft* (community) that accompanied his reign.

Ley's novel was different. Rather than promoting airships and rockets of the future as the means to restore the German empire, *Die Starfield Company* offered readers a love story of international cooperation. Set in the 1980s, the tale glorifies rocketry and aerial technology as the future of travel. The main character, Frank Daybor, is the German-born director of the Transcontinental, the West's largest airline. In order to combat a group of mysterious "air pirates," Frank teams up with Cora Samdarava of the Starfield Company, an India-based airline.

What follows is an interracial love story between Frank and Cora, as they share intimate moments in the sky. Through joint efforts, the European and the Indian grow to understand one another out of mutual respect. They fall deeply in love.

Although Frank is German by birth, his roots are incidental. He considers himself simply as a Westerner, in command of an internationally focused company. Cora is the most complex character. She has a difficult time with her mixed heritage. She is torn between a world of the past and a world of the future. As the "mistress" of the Starfield Company, Cora is both a traditional religious figure for her people and a modern pilot. Her wardrobe constantly changes from eastern and feminine silks to western and masculine flight gear. She is also struggling with two sides of herself and two camps of her people. One wing fights to modernize through technological and engineering might, while the other side or camp fights to preserve Indian tradition and identity. Cora struggles to find a balance.

By combining forces and sharing intelligence, the West and the East defeat the space pirates. The pirates are extraterrestrials whose intentions are hostile but unclear. The Starfield Company builds a rocket ship and launches missiles to destroy the "second moon," the base of the invaders. Their successful fight against the foreign menace has not only united Cora and Frank but has also united many nations of the world, which come together in solidarity. Cora and Frank's technologies have also shrunk the globe as more and more people feel at home in the air, free to lunch in Paris before sight-seeing in the tropics. After saving the world from the extraterrestrial threat, Frank and Cora marry. Their union represents the pinnacle of international and cross-cultural understanding by two powerful CEOs. Their space ship unites the world.

Ley wrote this novel as the rocketry fad peaked. It reflected his professional hopes and personal longings. It also reflected his naïve faith in scientific internationalism. He genuinely believed that science and technology could create a more perfect world. Engineers and aviators could save the world.

"Success, Failure, and Politics"

After the rocketry fad peaked in 1928 and 1929, public interest waned. Due to the onset of the Great Depression, many Germans had more immediate and everyday concerns. Ley recalled, "To say that things looked bleak by the end of 1929 is to understate matters. The Oberth rocket had failed to materialize. Winkler was forced to abandon publication of the monthly journal *Die Rakete*." Ley added, "Even the film [*Frau im Mond*] was only moderately successful." Despite the lavish premiere of the film, it was soon competing with "talkies." Ley remarked that the film had "dazzled us into confusion." Success had been fleeting.[5]

To make matters worse, 1929 marked the rise of Rudolf Nebel within the ranks of the VfR. Nebel had served as Oberth's first assistant on the publicity rocket stunt, which brought him prestige among Oberth's followers. Ley despised his tactics and personality. Historians have confirmed Ley's perceptions of Nebel as "more of a master manipulator and operator than an engineer." Yet he was also a veteran fighter pilot, likely skilled in machinery. Ley called him "a professed militarist" who lacked qualifications to work with rockets. Additionally, when Ley first met Nebel during a chance encounter, Ley was "dumbfounded." Nebel took a moment to brainstorm how a spaceflight society might be formed to further experimentation. Ley recalled: "I could not see any reason why he should want to compete with the VfR, but, on the other hand, he did not sound as if he did want to. I asked him outright and learned that he did not know of the existence of the society. Oberth . . . had never mentioned it." This level of disregard astonished Ley. He claimed: "What saved the situation was the fact that early in the same year Johannes Winkler had resigned as president of the VfR for personal reasons. Professor Oberth had become president and I vice-president." The presidency became an honorary title. Ley recalled, "I did most of the work." Ley managed the VfR in a donated Berlin office space. There was still hope for new publicity opportunities. For example, the group soon organized lectures and rocket displays, such as an April event in the public auditorium of the General Post Office.

Similar events happened during "Aviation Week" at Potsdamer Platz as well as within the basement of a large Berlin variety store.[6]

Then tragedy stuck with the death of popularizer Max Valier. When Valier experimented with liquid fuels in May 1930, his prototype motor exploded. A "steel splinter cut the aorta," Ley recounted. Valier "bled to death before anybody could do anything about it." The accidental death created a small public outcry against rocket experimentation. This outcry had been building momentum since the death of an adolescent boy who "was trying to build a large model of the Opel rocket car." This tragedy led to the introduction of an unsuccessful Reichstag bill aimed at banning rocket experimentation. Ley commented: "Valier's death was especially tragic in view of the fact that nothing had ever happened to him during all his dangerous and useless experiments with powder rockets. He died while engaged in his first really useful experiment, although the idea of mounting his motor in a car was, of course, ridiculous."[7]

Despite these setbacks and tragedies, there were still hopeful signs. In fact, Ley described the following years as a period of "success," before failures and politics got in the way. The VfR now included competent engineers, such as the eighteen-year-old Wernher von Braun. Ley had introduced von Braun into the group after the young man showed up at Ley's home. With the work of these assistants, attempts to perfect Oberth's "Kegelduse" culminated in a successful test firing on July 23, 1930. By September key members of the group were also testing newly designed "Mirak" rockets. Then Nebel furthered the goals of experimentation by leasing an abandoned army garrison in a northern suburb. Nebel dubbed the site *Raketenflugplatz Berlin* (Rocket Port Berlin). It was soon occupied by a ragtag group of unemployed and desperate engineers. Albert Einstein's son-in-law described these men as "officers living under military discipline. . . . They belonged exclusively to a world dominated by one single wholehearted idea." Ley remembered the site in less disciplined ways. He recounted improvised spaces, out-of-work engineers, and jovial comradery due to successful test flights of "Repulsor" rockets. According to Ley, the site produced 87 launches and more than 270 static tests.[8]

In the early days of Rocket Port Berlin, Ley was optimistic: "We did have a program of sorts, but while we knew precisely what we were *not* going to do, we could not formulate clearly what it was we were going to do." He added: "On the negative side we were certain that we would not touch solid fuels of any form. We also were not going to stick a rocket motor for liquid fuels on a car, railroad car, or glider. We were, in short, not going to do anything but build rockets." In a more direct account, he bluntly stated, "There was to be no nonsense about rocket cars." He also claimed that the overarching goal of early rocket experimentation, both in Germany and abroad, involved "honest and very serious attempts to solve purely scientific problems." He elaborated: "I happen to know with absolute certainty that they were not 'attempts to reach the moon.' Neither were they 'forerunners of transatlantic rocket airplanes.' And they were also not, as could be read occasionally in European newspapers, 'future deadly instruments of war.'" Ley further recalled: "It was exceedingly obvious . . . that one had to look upon a rocket as an embryonic spaceship."[9]

In spite of economic difficulties and hardships, the movement survived the early days of the Great Depression. Through the leadership of Ley and others the VfR continued, although its ranks diminished. Ley even proclaimed, "There is no 'rocket-scientist' in Europe, who is not a member of our 'Verein.'" Meanwhile, the establishment of the research site inaugurated the early days of experimental rocketry: "The scheme worked very nicely for one year." While Nebel negotiated for free supplies and discounts, Ley promoted the club and site in domestic and foreign publications. Engineers did the "real work." The "science" of rocketry progressed.[10]

However, Ley's patience with Nebel and the group diminished day by day. As the de facto leader, Nebel could be dishonest with the public, donors, and organizations. For example, Ley recalled: "Nebel promised that man-carrying rockets could be built on short notice and he began a discussion of the theory of the station in space with the words: 'Judging from our recent experience in this matter . . .'" When Ley told Nebel that his claims could not be substantiated, Nebel responded, "That doesn't matter, advertising and science are two different things.

I'm a specialist in successful salesmanship." Ley believed that Nebel's success with this approach was always short-lived: "He did get a lot out of people in the way of donations, but he never got anything more than once." The group needed a long-term strategy. They needed to be honest with the public. Ley also grew disturbed by the increasing militarization of the site. When Oberth resigned his presidency, Major Hans-Wolf von Dickhuth-Harrach accepted the job. Ley would eventually write about Dickhuth-Harrach in neutral terms. At the time of his appointment, Ley grew anxious about the militarization of the agenda. He later recalled that Dickhuth-Harrach's appointment "produced all the groundwork for a psychological explosion which I postponed as long as possible."[11]

Nevertheless, Ley was pleased by the work of the VfR and the successes of the Rocket Port. He concluded, "The year 1931 brought real progress." Much of this progress culminated in a newsreel that showcased rocket experimentation. According to Ley's notes, there were 23 demonstrations for clubs and other societies together with 9 simply "for publicity." Ley remembered, "Everything, or most everything, went fine. . . . Our finances were all right. Most members paid their dues. . . . Various public meetings and lectures brought in some more money and the demonstrations helped greatly." All the while, engineers made real progress, particularly when it came to the "Repulsor" test flights. Ley fondly recalled the role of the Rocket Port in providing "shelter, food and a little pocket money to unemployed mechanics." Socialists and pacifists were welcomed at the site, in spite of Nebel's politics and his "loud and accented voice, surprisingly much like Hitler's." Nebel was in fact a Bavarian, from the same border region as Hitler.[12]

During 1931 Ley's international correspondence also increased, as he openly shared information with British and American rocket enthusiasts. In particular he corresponded quite often with G. Edward Pendray of the American Interplanetary Society. When Pendray visited the Rocket Port in April 1931, Ley served as his host. According to Ley, they got along well, and Pendray reported on the group's activities and accomplishment in the club's journal, *Astronautics*. Ley also wrote several articles for the journal. Ley's meeting and correspondence with

Pendray marked the beginning of a friendship. Ley openly shared technical information. While Goddard and Oberth worked in secrecy, Ley outlined the details of tests on postcards. His correspondence reflected his general attitude about international cooperation among a field of pioneers. In a comment about Goddard that probably irked Pendray, Ley wrote, "I don't think very high on [sic] the works of Goddard, you know. Not because he is one of your own men, but because he always has his 'secrets.' I have always learned, men with secrets have no secrets,—his last plan for a rocket-plane was pure nonsense. Bluff, only!" Secrets signaled weakness, not strength.[13]

Ley also attempted to use Pendray as a key American contact. In 1931 Ley asked Pendray for a copy of Goddard's *A Method of Reaching Extreme Altitudes*. "I have read the book, but the copy belonged to the Bibliothek of the University," he claimed. He also asked Pendray to relay his novel, *Die Starfield Company*, to a Gernsback publication. On a personal note he added, "Don't fear for our safety. The communist one-act-plays are only in certain streets in the slums and only for one or two days." Outside the confines of the Rocket Port, politics could turn bloody and riotous on both ends of the political spectrum.[14]

These personal letters increased in the coming months. For example, Ley gave Pendray a long-winded explanation for the tensions between the VfR and Oberth. According to Ley, Oberth intended continuing his experiments in a secretive fashion in his own country. He also "doesn't like to be President of a society in another country and that is true! We finally think the same." Ley expressed his optimism: "Now we are working without connection to him and we are sure, we shall do good work." During this time Pendray sent Ley many different English articles as well as science fiction stories. Ley sincerely thanked him for helping to improve his English reading ability. Ley also thanked him for news of a replacement copy of David Lasser's *The Conquest of Space* (1931). As the first English book on rockets, it advocated for solely peaceful uses of the technology.[15]

The general situation soon deteriorated at the Rocket Port. When a rocket crashed into a police building, "any further experimentation was forbidden then and there." Fortunately the police lifted the ban under new conditions. Experiments continued. However, with worsening economic conditions, membership in the VfR declined while sources of funding diminished. A key investor could no longer support the group. In a last-ditch effort for publicity and funding, the VfR Board of Directors agreed to plan and advertise a manned rocket flight. Ley vaguely disguised his protest, telling readers, "Nebel spoke about the next plans of the VfR. Because the VfR wants to do scientific work, it needs money. But nobody is so much interested to spend even smaller sums for this purpose, they all want to see sensational rocket-shots." In a revealing statement about the VfR's connection to the later and infamous "Magdeburg Project," Ley wrote, "The Vorstand of the society agreed under these circumstances to do ONCE a real show—work and build the first manned rocket for liquid propellants." This project would end in disgrace and legal repercussions. Ley would eventually deny all connections between the VfR and the Magdeburg Project.[16]

In his memoirs Ley lamented the deteriorating situation of 1932 and early 1933:

> The newsreel and our victory over the police were our last triumphs. What followed afterward was a hopeless struggle against political tension and economic misery. It was a hard winter climatically. And it was the fatal winter under Chancellor Bruening when Adolf Hitler suddenly assumed prominence. It was a winter during which the roster of VfR members shrank to less than three hundred, most of them deprived of their livelihood. It was a winter during which there came many letters saying that no further dues would be forthcoming because "all money belongs to the Führer."[17]

Ley claimed, "The general situation was deteriorating from day to day. Nerves were frayed; practically every meeting of the Board of Directors . . . led to violent clashes, caused by, in the last analysis, political difference. The deterioration was rapid." Rudolf Nebel became intolerable, according to Ley. He "attacked everything and everybody." He openly criticized Ley for sharing information with foreigners. "Speaking with a careful imitation of Hitler's mannerisms," Ley recounted, "he declared that he would leave the VfR to die and join the army." Nebel is also reported to have claimed, "If we say it can work, we'll get army money to try it."[18]

Back in December 1931, Ley had left Berlin. He later claimed, "In order to raise money I went on a lecture tour in East Prussia." Ley gave readers the impression that his trip was a virtual blitzkrieg of publicity. According to a letter to Pendray, his trip was much longer. Ley also later wrote, "My lectures were not spaced very closely." For the most part, Ley's "lecture tour" was an extended vacation with friends and family, interrupted by occasional efforts to raise money for the VfR. What can be verified is Ley's growing detachment from the inner circles of experimenters. The scene became secretive. Politics got in the way of progress, in his view.[19]

The Militarization of Rocketry

According to Ley, Rudolf Nebel had militarized the agenda. His collaboration with the German Army, Ley remembered, "in retrospect looks like pure comedy." After Nebel failed to persuade an investor, he wrote a "technically wholly inadequate and senseless 'Confidential Memo on Long-Range Rocket Artillery.'" Soon afterwards, a "burglary attempt" indicated that someone was closely monitoring the site. Ley recalled: "And then I saw that we were no longer 'in all the rooms.' Somebody else was there: the busily plotting German army. Suddenly they supervised everything, unseen, but efficiently and some of Oberth's and Nebel's claims worked nicely into their hands." Ley may have been a curious onlooker as plain-clothed generals visited the site. In reality, these officials were skeptical of war rockets and Nebel's claims. Ley grew paranoid and suspicious.[20]

Then the German Army ordered a rocket demonstration at a proving ground at Kummersdorf. An "ultranationalist" colonel of the Army Ordnance Office was interested in rockets. According to Ley, Nebel did not notify the VfR's Board of Directors, and he took engineers Riedel and von Braun to the site of the demonstration. Although the demonstration was not a success, according to Ley, "A month or so later the Army hired Wernher von Braun, who disappeared from our view for a while." At the time, Ley did not know the details. The young von Braun simply disappeared. Meanwhile, Nebel continued to antagonize the German Army, due to his dual attempts to use their resources for secret experimentation while generating publicity for those same experiments. These clashes led to a showdown after the Nazi seizure of power.[21]

Ley remembered, "Everything collapsed at once." He would later characterize this period as "the beginning of modern rocketry," due to the activities of engineers Dornberger, von Braun, and others. However, at the time Ley mourned the death of his field of expertise. In his perspective Nebel had destroyed the VfR. He had also betrayed the cause by militarizing the rocket. "The program of the VfR," Ley claimed, "had been entirely different, it had aimed at the creation of the spaceship as the ultimate goal." Ley grew intolerant of the Rocket Port. He wanted nothing to do with Nebel, the German Army, or military rockets. Simultaneously, key members of the group voiced their ongoing displeasure with Ley's international correspondence. He recalled, "I became known as a 'xenophile,' a man who keeps up correspondence with foreigners." The atmosphere became secretive as rocketry became militarized.[22]

Ley tried to fight these developments. In fact, during the early half of 1932 he had spent much time in Berlin libraries, researching ballistics, trajectories, and weapons designs. He aimed to debunk scientifically the war rocket as promoted by Oberth and Nebel. He recalled, "Both had talked and written a lot about war rockets, and many people had believed." Ley instinctively distrusted their self-serving and opportunistic claims. His "long study" convinced him that "rockets in battle can never be as efficient as guns in battle . . . [while] the bombing airplane can carry an immensely superior load." This conviction stayed with Ley for a decade. The ineffectiveness of the war rocket was also an obvious lesson of history. Ley outlined that history in his sixteen-page *Grudriß einer Geshichte der Rakete* (*Outline of the History of the Rocket*, 1932).[23]

As Ley grew disturbed by the presence of the army, he also grew increasingly frightened by the domestic and international scene. For example, in the late spring he opened two copies of the latest editions of *Wonder Stories*. These issues published "The Final War" by Carl W. Spohr. The novel characterized Western civilization as relentlessly addicted to war at all costs. The novel distressed Ley profoundly. He wrote to *Wonder Stories*, claiming, "I didn't want to write again so soon. I must! I must!" He had spent all day reading the novel. He wrote, "Unnecessary to say it's the best of all war stories . . . written by a man who saw the hell of the Great War." He asked: "Must final war come? Must mankind wipe itself out?"[24]

In a stunning contradiction to his earlier conclusions about war rockets, Ley wondered, "Maybe some new inventions, like air raids by rockets spreading death over a whole country, will make the biggest danger smaller." Ley then directly labeled "the owners of factories of guns and ammunition" as war profiteers who would benefit from mass hysteria. He concluded with a hopeful yet profoundly naïve comment: "And if men of one nation learn and see enough of other nations, they will lose the idea of war against a nation in whom they have friends . . . there is a hope." That hope quickly faded by the time that Ley met science fiction editor Hugo Gernsback in June 1932. Even the "father" of pulp science fiction expressed his urgency. Ley informed Pendray, "He told me . . . Men, hurry, you are the hope of the world and if you are down long enough, Bolshewism [*sic*] will eat us all together!" "Isms" were conquering the world.[25]

Soon, Ley lost all hope for both his nation and his field of expertise. After Hitler became chancellor in January 1933, the Reichstag burned in February. Authorities blamed a communist for the fire, and mass arrests followed. Certain publications were immediately banned. In late February the Reichstag Fire Decree posted new restrictions for all Berliners. The document suspended a long list of civil liberties, including the freedoms of expression, assembly, and the press. The document also suspended the privacy of postal, telegraphic, and telephonic communications. Then the Enabling Act of March 1933 granted Hitler dictatorial powers. It also announced that all new laws would be printed in the *Reich Gazette*. The Gestapo was established in April, and its growing networks of spies and informants must have seemed conspicuous. On April 7, 1933, the "Law for the Restoration of the Professional Civil Service" outlined the immediate purge of all enemies of the state from civil service. Soon scientists and researchers became key targets of the state. Simultaneously, in April 1933, the German Student Association demanded "Action against the Un-German Spirit," which led to massive book burnings at major cities and university campuses. Among many other books, officials burned Lasswitz's *On Two Planets*. The Nazis burned Ley's favorite science fiction novel.[26]

During that same month rocket enthusiast and engineer Rolf Engel and a colleague were arrested and charged with "negligent high treason" for corresponding with foreign engineers. The documents of the group were confiscated, which indicated that certain army officials wanted to eliminate experimentation and public discussion. Rudolf Nebel and key members of his group were under similar scrutiny, particularly during and after the bizarre Magdeburg Project. To sum up, Nebel promised to launch a manned rocket, with the goal of proving the "Hollow Earth Theory." An engineer convinced the city of Magdeburg to pay the bill.

According to most accounts, when Dickhuth-Harrach and Ley learned of the plan and the corresponding publicity for the launch, they drew up a list of charges against Nebel. They aimed to expel him from the VfR. The debacle created a final schism between Ley and Nebel. Ley was not simply outraged by the implausibility of the stunt. He was outraged by the pseudoscientific theories behind it. It was one thing for a pioneer like Valier to believe in glacial cosmogony. It was another matter when such nonsense affected the agenda and reputation of the VfR. This pseudoscience undermined the scientific cause. It associated spaceflight with cranks. It was far more dangerous to the cause than publicity stunts with powder rockets attached to automobiles. Most likely Ley also considered the affair as a personal insult. By 1934 Ley and Dickhuth-Harrach *were* the VfR. When Nebel associated the Magdeburg Project with his organization, Ley had to respond.[27]

Later memoirs recounted a series of events that are difficult to verify. Allegedly, Ley and Dickhuth-Harrach took Nebel to court. They sought to expose Nebel's misuse of funding as well as his questionable tactics. To their dismay the district attorney dropped the case for lack of evidence. Ley explained, "The District Attorney, seeing that Nebel wore a swastika armlet, was afraid to act . . . the ensuing conversation ran something like this: 'Herr Major, I hesitate to do anything . . . I have noticed that he wears a Party Armband [the District Attorney did not]. . . . These are revolutionary times.'"[28]

Apparently Nebel's fascist ties saved his skin. Yet he was on thin ice, as the army continued to consolidate control of rocketry. In fact, his continued correspondence to England resulted in a Gestapo raid on the Rocket Port. From Ley's perspective, the investigation of the site appeared quite mysterious: "On one occasion I was not permitted to enter, being told by one man with some insignia of rank on his collar tabs that the Gestapo was there to seize documents and equipment." Ley later recalled his perception of Nazi officers suddenly intruding into the spaces of Rocket Port Berlin: "As far as we were concerned the 'reborn Germany' consisted of two or three meticulously booted and uniformed young men who gave the impression of being homosexuals. Being of the ripe old age of nineteen or thereabouts they carefully patterned their speech after the Führer's, unless they grew excited and forgot to do it." In later memoirs Ley removed the derogatory reference to homosexuality. This episode marked the end of both the VfR and Rocket Port Berlin. Ley recalled: "It seemed as if there were no way out, with everybody's hands tightly tied by a ruthless totalitarian regime." In Ley's perspective, progress on rockets stopped. The key organization had been disgraced. The key site of experimentation had been closed. Ley's friends and colleagues became more secretive. At the time, Ley was unaware of the true extent of the army's consolidation of rocket engineers, which was minimal prior to 1937.[29]

Ley did not immediately give up. He and Dickhuth-Harrach resigned from the VfR Board of Directors. Ley wrote of his intentions of collecting VfR members to migrate to a new organization. By February Ley claimed success by transferring VfR members into the E. V. Fortschriftliche Verkehrstechnik (EVFV, Society for Progressive Transport Technology). Although "it was a difficult and not very pleasant job to clean up the whole mess," Ley claimed that he succeeded in isolating Nebel and effectively disbanding the VfR. While planning "to make propaganda again" with a new journal, Ley identified himself as "the leading spirit" of the original society. He announced: "We are of the opinion that the ideals and the good old tradition of the VfR shall not be allowed to perish." Privately Ley told Pendray, "Nobody is left of the Nebel crowd. His name means mist or fog . . . and that's what he is and what he does always."[30]

At the moment when international correspondence became a dangerous activity, Ley kept writing. In fact, he pleaded for greater international cooperation between the (renamed) American Rocket Society and his new organization. He also spread the word about the newly formed British Interplanetary Society, led by chemist Phil E. Cleator. In January 1934 Cleator had visited Berlin, where he spent two days with Ley. Most likely Ley openly shared information with the British scientist. Despite the arrests and the surveillance of the scene, Ley also continued to correspond with Pendray about rocket fuels, motor designs, and cooling systems. They considered combining organizations into a single international rocket society.

It was a dangerous time for a scientific internationalist. The state moved to silence all publicity. Amid the constant redressing of Berlin and its citizens to conform to the designs of the state, Josef Goebbels's Propaganda Ministry issued a decree on April 6. This decree banned discussions of the military uses of rockets as well as the publication of their technical details. As historian Neufeld describes, "From the standpoint of the public, rocketry disappeared in 1934 because of the imposition of censorship." Ley had been the publicist of a spaceflight movement. He was now out of a job. Talk of rockets was forbidden.[31]

Ley began to plan his escape. The earliest hint came in an autobiographical letter to Pendray. Ley briefly summarized his contributions to the scene of rocketry, stating, "It seems to me that I'm the best source of information in the world." He added the following comment about himself: "Not married, not engaged, want to see the world especially England and America, interested still in movies, like Joan Crawford, don't like Mae West. . . . Is it enough? [What] I look like you know, it could be blonder for the time being (don't mention the last!)" In decades to come this last statement would fuel rumors of Ley's Jewish identity. The rumors are baseless.[32]

Following this correspondence, police arrested Nebel. He had written a pamphlet called *Rocket Torpedo*. Nebel sent his pamphlet to the paramilitary organization of the Nazi Party on the verge of the Night of Long Knives, when the Nazi regime committed a series of political murders. After the army denounced him to the Gestapo, a sympathetic figure intervened. Although Nebel escaped the ordeal, the arrest must have frightened Ley further. Another spaceflight enthusiast named Werner Brügel had already published a profile of key rocket men in 1933, prior to the ban. In August 1934 Brügel planned to give a radio talk on the use of rockets for exploration of the stratosphere. The Gestapo raided his residence and confiscated his documents. Although it was an arrest of a minor player, it illustrated the Army Ordnance campaign to consolidate rocket development. Meanwhile, Ley continued his foreign correspondence. By late 1934 he suspected that someone was monitoring his mail.[33]

Pseudoscience in Naziland

Years before Ley entered the scene of rocketry, reactionary politics had left its mark on the scientific community. As early as 1921 the attack on "Jewish physics" was well under way. Physicist Johannes Stark attacked Einstein, arguing that he had "betrayed Germany and German science with his internationalism," along with his supporters' tendency to publicize scientific theories through foreign lectures. By 1933 Einstein had become the most infamous symbol for "the 'internationalist' influence which Hitler's movement was determined to eradicate." After the Nazi seizure of power, the physics community by no means fully agreed with this sentiment. In fact, many were troubled by the total coordination of German society. Others, particularly doctors and anthropologists, easily adapted. It was not long before scientific intellectuals began to write textbooks like Philipp Lenard's *Deutsche Physik*, which claimed that everything created by men, including science, could be attributed to blood and race. By 1935 many scientists had fled, mobilized for the state, or simply tried to co-exist. The internationalists became enemies of the state.[34]

Scholars have long documented the rise of pseudoscience in Nazi Germany, which is an outdated narrative. Many have commented on a lethal mixture of anti-Semitism, mystical philosophy, and racialized eugenics that flourished unchecked. At the moment when the army consolidated rocket researchers, Himmler was trying to establish the *Ahnenerbe* (Ancestral Heritage), which was a research society focused on holistic science. The organization sought to eliminate the distinctions between natural sciences and the arts, while promoting holistic worldviews that conformed with Nazi ideology. Nazi scientists, in turn, were often mixing archaeology and anthropology with mythology, astrology, and the occult.[35]

These movements could also be grassroots. Ley perceived a dramatic increase in the popularity of certain doctrines that had once occupied the fringes of science. In a later article he described the rise of "pseudoscience in Naziland." In general he blamed irrationalism, mysticism, and anti-intellectualism: "When things get so tough that there seems to be no way out," he joked, "the Russian embraces the vodka bottle, the Frenchman a woman, and the American the Bible." He continued:

> The German tends to resort to magic, to some nonsensical belief which he tries to validate by way of hysterics and physical force.... It was the willingness of a noticeable proportion of the Germans to rate rhetoric above research and intuition above knowledge, that brought to power a political party which was frankly and loudly anti-intellectual.[36]

"Small wonder," Ley added, "the pseudoscientists experienced a heyday under such a regime." In Ley's perspective, the pseudoscientists had existed for many years, struggling to achieve some degree of respectability. Now they flourished amid the broader embrace of irrational politics and the vulgarization of holism. Of particular offense to Ley was the popularity of the "Hollow Earth Doctrine" and the "World Ice Theory." These were not legitimate areas of scientific speculation. Instead, they involved "dream-reasoning fitted into the Nazi philosophy." Pseudoscientists like Hans Hörbiger "literally had millions of fanatical supporters who would interrupt educational meetings with concerted yelling, 'Out With Astronomical Orthodoxy, Give US Hörbiger.'" His followers seemed to privilege mystical intuition over empirical knowledge. Ley compared other movements to cults with fanatical and obedient followers.

In Ley's view the pseudoscientists not only had conformist followers but also displayed a profound intolerance for dissent. In a letter to

Ley, Hörbiger allegedly wrote: "Either you believe me and learn, or you must be treated as an enemy." For Ley it was even more amazing that the public believed such nonsense, when Hörbiger's publications and letters "revealed clearly that he was not even a good engineer," let alone a decent astronomical theorist. Unlike a legitimate scientist, he based his theories on intuition and visions. Any educated German could "pick flaws in this theory . . . as easy—and as pleasant—as gathering Japanese beetles from an infested flowerbed." Yet in Ley's memories much of the German public embraced the pseudoscientific ideas, as if they represented a new gospel. Then, within a "powerful popular movement in pseudo-intellectual circles . . . adherents declared threateningly that now everybody MUST believe Hörbiger, or else." The pseudoscientists had become rigid and closed-minded authoritarians. They sought to impose their magical thinking on non-believers. Citizens would be converted, or else.[37]

These perspectives should make historians cringe. Ley conveniently ignored the mainstream history of American and British eugenics, which greatly influenced Nazi policies. He ignored the long legacy of scientific racism in medicine, anthropology, and psychology. His distinction between real science and fake pseudoscience does not hold up to scrutiny. It also resembles other narratives that explain Nazi Germany as an aberration or deviation in Western scientific progress. Historians can easily distance themselves from these perspectives. Nevertheless, Ley viewed his world through these distinctions. He associated Nazism with propaganda, a paranoid state, and the spread of irrationalism. His fellow Germans, in his opinion, supported a mystical, dogmatic, and unreasoning "pseudoscience in Naziland." Simultaneously, they embraced a mystical and anti-intellectual style of politics. The two worldviews overlapped, in his judgment. A mental fog had spread throughout Germany, taking possession of rational minds and a culture that prided itself on its technological and scientific might. Germany had grown spellbound with irrational delusions. There was no place for sanity or rational science. Progress stalled.

The Escape

Following the brief arrest of popularizer Werner Brügel, P. E. Cleator of the British Interplanetary Society received "a rather mysterious communication." An unnamed friend of Ley's smuggled a message out of Germany. Writing from Holland, the informant discussed Ley's political problems. Cleator summarized the information for Pendray: "Apparently there is some trouble brewing in Germany . . . trouble about which Herr Ley dare not write." Cleator claimed that Ley's mail was being opened and examined. Ley asked Cleator not to use any official stationery or envelopes of the British Interplanetary Society. He pleaded for Cleator to pass along the message to the American Rocket Society. Cleator noted, "'Rocket' is taboo." Then Cleator received a different letter directly from Ley. Cleator relayed Ley's inquiry, asking Pendray "if you can think of anyway [sic] in which he could earn some money during his stay in America."[38]

After receiving this information, Pendray wrote to Ley on blank stationery: "It occurs to me that you might have some time this winter to visit America. . . . We have ample room to keep you for an indefinite time and I could think of no greater pleasure than to serve as your host in America." Five days later Pendray wrote to the American Consulate in Berlin, asking that they grant Ley a visa. Soon he received a response to an earlier inquiry with the National Council of Jewish Women, which informed him, "If your friend has funds . . . and if he secures a visa from the American Consul, he should have no difficulty in getting here." Pendray wrote an affidavit swearing to be financially and legally responsible for Ley. In a cover letter Pendray also argued that Ley was "a good friend . . . the moving spirit of rocket experiments and research in Germany and despite his youth . . . a man of considerable linguistic scientific and literary achievement." He added, "I am prepared to share my home with him and to provide him with food, clothing and necessary expenses until he can establish himself in this country."[39]

Meanwhile, Ley corresponded with Cleator, attempting to arrange a time and place for his "vacation" to Great Britain, where he would board a ship headed for the United States. In the letters to Pendray, Cleator seemed quite confused by the affair, due to Ley's guarded and cryptic language. Cleator told Pendray, "As I understand the matter at present, his leaving Germany is nothing more than a holiday." Pendray insisted otherwise. After the exchange of more letters, Ley told Pendray: "You are absolutely right and he is equally wrong . . . I'll never forget what you have done for me. I hope that some day [sic] I'll be able to show you how grateful I am."[40]

Ley disguised his "vacation" as a journalistic trip. This may have allowed him to carry several orders from different editors. He also made arrangements for a trunk filled with books to travel separately to Pendray. According to the original plan, Cleator would visit Germany in early January, and Ley would accompany him on his return trip to Great Britain. Unfortunately for Ley, he had problems obtaining a permanent visa. In a telegram Ley pleaded with Pendray for a deposit of 500 marks. After Pendray sent a cablegram to the American Consular Service in Berlin, he received the following reply: "An immigration visa cannot be granted to Mr. Ley because of his serious physical defect (he is practically blind in one eye) and also because he has practically no personal resources." Had the situation not been desperate, one could imagine Ley joking about being neither blond enough for the Germans nor eagle-eyed enough for the Americans. Nevertheless, he remained calm, working with the consulate to obtain a renewable one-year visitor's visa. Ley obtained a visitor's visa, but it remained to be seen whether he could easily leave the country. On January 30, 1935, Ley told Pendray, "Now everything is O.K. I've got my visa and I'll get my tickets tomorrow. I'll leave Berlin Sunday next [February 3rd] and go to London first." He then stated, "An old dream of mine becomes true with this trip and I have you to thank for it . . . I'll try to cause as less trouble as possible in your house."[41]

On February 3, 1935, Ley took a train from Berlin to Düsseldorf before crossing from Hook of Holland to the English port of Harwich. He had no difficulty crossing the German border. He later recalled, "I could have taken anything I wanted past that guard. He didn't even search me—just checked to see if my name was on their black list. It wasn't so he let me by and wished me a good trip." As noted by historians Sharpe and Ordway, with his departure, "the flow of rocket society news [out of Germany] virtually ceased—and hardly anyone noticed."[42]

After spending a few days in London with the editor of *Armchair Science*, Ley made the journey to Liverpool to stay with Cleator. Together they sat down to inform Pendray of the news. Cleator summarized the situation: "It will suffice of I [to] say that rocketry (experimentally) is virtually banned in Germany. Brugel [sic] is in a concentration camp, and Zucker has been put in prison. But Willy has got here!" The statement on Brügel was not true. According the Cleator, Ley had a total of "10 marks in his pocket (about 2½ dollars)." Ley then clarified in his own words: "Things are about as Phil told you, but it is not so bad as it sounds. Please, don't get the idea to publish anything of it, my relations in Germany would have serious trouble . . . I'll tell you everything personally."[43]

On the day before Ley's departure, Cleator complained to Pendray in a section of a long letter titled "Willy (or Won't He?)." Not only had Ley constantly annoyed Cleator by being late, but he also took advantage of Cleator's generosity. Cleator claimed, "Willy does not appear to have the slightest idea of the value of money—or at least of other people's." Cleator compared this rudeness to a characteristic of the German "race." He stated, "They seem to take most things entirely for granted. Other nations, it would seem, exist to run around them. . . . Well, I've done all the running around I want to do for a bit!" Cleator then downplayed the urgency of Ley's escape, claiming that he was perfectly safe to continue theoretical work on rocketry. "Willy's only reason for leaving Germany, therefore, is that he wants to experiment. Well, that seems reasonable enough, except that we both seem to have been misled over the whole business." Cleator then expressed his fear that Ley would be an unproductive burden on Pendray. "My only hope," Cleator wrote, "is that he does not prove to be so helpless in America as he has, of necessity, been here. . . . Willy will leave tomorrow from Southampton aboard the Olympic. And in some ways, I must confess, I can't honestly say I'll be sorry."[44]

Nonsense in Naziland

One word characterized Ley's view of the scientific and political scene: nonsense. In his perception of events, he had watched many Germans embrace a nonsensical and anti-intellectual regime. He witnessed fellow rocket enthusiasts make nonsensical plans to launch a manned rocket flight to confirm a pseudoscientific theory. He then perceived the transition of a scientific field from an open and honest forum of exchange into a secretive world of military oversight, with its corresponding and nonsensical hopes for war rockets. In his perspective the state had essentially killed the field of rocketry. There was no future for the rocket as a wonder weapon. Likewise, there was no future for a scientific society in Germany. The totalitarian state had clamped down on progress. The republic of letters was silenced.

What had begun as a combination of publicity and media had transitioned to a phase of experimentation. That phase now ended in failure and politics. Open borders were forcefully closed. Dogmatic conformism replaced the freedom of inquiry. Science and technology had become tools of a paranoid and authoritarian state. The field of rocketry had been smothered in a blanket of secrecy and censorship, just as cosmopolitan Berlin had been redressed in banners and propaganda. Under such conditions, there was simply no way forward. Freethinking became dangerous. Imagination became confined to nationalistic fantasies. Thus the fascists had destroyed the very engine that drove science and technology forward. Needless to say, many of these perceptions did not match reality. Yet they became the threads of a larger narrative in Ley's mind. As we will see in later chapters, many other émigrés shared his views on the death of a science in Germany.

The *Olympic* sailed for New York on February 14, 1935. It was the last voyage of the luxury liner. Ley recalled, "She was big and beautiful, but too old." Such a vessel was obsolete and far too slow. During the voyage Ley must have reflected on his family and friends. It is not clear if he said his goodbyes. As he saw the European coast fade into the distance, he must also have reflected on his homeland and what it had become. He had loved Berlin. He had made a name for himself as a science writer and publicist. He had befriended important people, such as Hermann Oberth and Fritz Lang. He had also contributed to an active period of rocket experimentation. Yet everything had collapsed. Ley hoped for a new start in the United States. Perhaps the scene would be different in the land of aspiring engineers. Never would Ley return to Germany. He was twenty-eight years old.[45]

FIGURE 1. Ley as an aspiring rocket engineer, circa 1936. Courtesy of the Smithsonian National Air and Space Museum, NASM 77-6019, box 9, folder 10, WLC.

FIGURE 2. Publicity photograph for *Days of Creation* (1941), painting by Olga Ley. Courtesy of the Smithsonian National Air and Space Museum, A-4787-C, box 9, folder 11, WLC.

FIGURE 3. Ley publicity photograph, possibly used to promote *Days of Creation* (1941). Courtesy of Xenia Ley Parker.

Right: FIGURE 4. Willy Ley and Olga Ley, self-portrait, circa early 1940s. Courtesy of Xenia Ley Parker.

Below: FIGURE 5. Ley at book signing event for *Conquest of Space* (1949). Courtesy of Xenia Ley Parker.

Left: FIGURE 6. Ley and daughter Xenia, circa 1951. Courtesy of Xenia Ley Parker.

Below: FIGURE 7. Ley and daughters Sandra and Xenia, circa 1949. Courtesy of Xenia Ley Parker.

FIGURE 8. Ley posing in nature, date unknown. Courtesy of the Smithsonian
National Air and Space Museum, A-4791, box 9, folder 11, WLC.

FIGURE 9. Ley with Fourth of July sparkler, date unknown. Courtesy of the Smithsonian National Air and Space Museum, A-4792-A, box 9, folder 11, WLC.

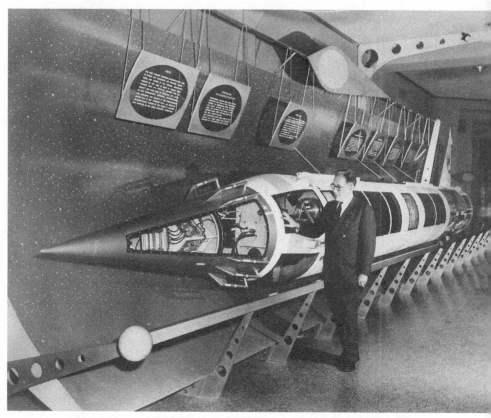

FIGURE 10. Ley publicity photograph, possibly taken at the Abraham & Sons department store exhibition, circa early 1958. Courtesy of Xenia Ley Parker.

Adventures of a Romantic Naturalist

By early February 1936 Willy Ley had been in the United States for
one year, and he stood ready to launch a rocket plane. Ideally the vehicle would soar 400 yards from Greenwood Lake, New York, to cross
the New Jersey state line. The rocket plane would be launched from a
catapult at an angle of 23 degrees. The rocket motor would maintain
thrust for 30 seconds. If successful, Ley would take credit for the first
rocket plane flight in the United States. It might later be hailed as "the
beginning," he recalled.[1]

Ley hoped that this demonstration would generate interest in rocketry, which in his view was sorely lacking in the United States. He later
expressed his surprise at this situation. In spite of a flourishing niche
market for science fiction pulps, "the idea of spaceflight was by no
means popular yet, especially in the United States." There had been
few concerted efforts to excite audiences about a future of Americans
in space. Ley recalled that scientists like Goddard "were told to take
their science fiction plots home with them." Rockets were the stuff
of *Buck Rogers* and *Flash Gordon*. With the flight of the first American
mail rocket, Ley hoped to change the situation. Perhaps his Weimar
tactics would work in a new setting.[2]

Ley also hoped to improve his situation. His first year in the United
States had not been easy. When he arrived in February 1935 he went

to Jones Beach to study the horseshoe crab. One could imagine him marveling at the creature, while pondering his situation as an impoverished refugee, homeless and jobless. Luckily, he had the support of G. Edward Pendray of the American Rocket Society (ARS). Ley lived at Pendray's home in Crestwood, New York for five months, where he witnessed or participated in experiments. Ley also contributed to the ARS journal as well as other publications.[3]

Ley owed his life in America to Pendray. However, Ley soon distanced himself from both Pendray and the ARS. There may have been a personal falling out. Regarding the ARS, Ley later stated that they simply imitated the German Rocket Society. Although the ARS progressed to the experimentation phase, a "financial vicious circle . . . caught it even more rapidly than it caught the German Society." He concluded: "I have to state, however, that I do not believe that a Society, unless backed by a few wealthy and very generous members, has a chance to progress much further than the German Rocket Society did." The ARS would otherwise need dues payments from 20,000 members when it seemed there were "not 20,000 people in the world that know enough about rockets and think enough of the importance of rocket research to support such a society for a number of years." By 1943 he was convinced that the ARS was rather pointless: "The more time I have had to think about it the more have I arrived at the conclusion that the VfR progressed as far as any society can progress. . . . Experimentation had reached a state where continuation would have been too expensive for any organization, except a millionaires' club."[4]

In 1935 Ley may have voiced such views, which would have angered Pendray. Likewise, Pendray's lack of enthusiasm for spaceflight probably angered Ley. Pendray was more enthusiastic about rocketry's impact on aeronautics and transoceanic flights. By June Ley was staying elsewhere. In a somewhat formal letter, Ley informed Pendray of his departure from Crestwood to stay with the van Dresser family in New York City. Ley lived with the van Dresser family for a few days, before moving into a boarding house. He also informed Pendray of his difficulties in finding office work. Ley's tone was apologetic and thankful: "But I will not miss the opportunity of this letter to thank you very much for all you have done for me . . . I will never forget it." For the meantime, Ley kept some ties to the ARS.[5]

Whereas the ARS could not further the cause, an investor could. Ley had begun his search in April, and he found a stamp dealer named Frido W. Kessler. Ley persuaded Kessler to invest in public demonstrations of rockets. Pendray explained Ley's plans: "If his plan goes through Ley will be the chief engineer of the project, which as it is now lining up will be a most ambitious and worthy one. Needless to say, the whole business is now confidential." Ley and Kessler made a three-year agreement to conduct public experiments with mail-carrying rockets. While Ley would design and build rockets, Kessler would control the publicity of the spectacles as well as the production of stamps and postcards. The costs of these launches would be offset by selling souvenirs after they had been mailed by rocket. The first launch was scheduled for February 9, 1936. Kessler's newly formed Rocket Airplane Corporation of America sponsored the flight.[6]

Proving Ground

Ley's mail-carrying rocket plane was dubbed *Gloria*. On the day of the first flight Ley and Kessler revealed *Gloria* to the public, which consisted of a few journalists and a crowd of five hundred onlookers assembled at Greenwood Lake. The crowd endured the bitter cold for several hours. Aviation writer Stan Solomon later described *Gloria* for the readers of *Air & Space* magazine: "Except for the graceful curve of the lower fuselage, the craft had not the slightest suggestion of streamlining. It resembled a giant version of a crude free-flight model airplane." A reporter at the time called the plane a crude "flying fish with its flat belly." Most of the materials used in the construction of *Gloria* were bought at a local hardware store. Despite its amateur-looking design, it contained "the most powerful explosive mixture known to man," according to the press. It also carried 6,000 letters and postcards.[7]

Ley had prepared for months. He had made numerous experiments with the liquid-fueled motor, and initial tests "proved to be worthwhile," he recalled. Not only did Ley's motor fire consistently on a test stand, but all plans for the flight had been "nicely calculated, checked and rechecked." The weather was the most notable obstacle, after blizzards caused several delays. Ley was soon ready to dazzle the public. He hoped to establish himself as a promising rocket engineer. His future prospects rested on the outcome. His reputation was at stake.[8]

After the police pushed the crowd to a safe distance, other factors delayed the launch, until Kessler yelled, "Let it go." Dressed in a crude safety suit, Ley approached *Gloria* with a small torch to light the rocket exhaust. He then jumped back, waiting for the rocket to soar into the air with the assistance of an angled catapult. The rocket performed poorly, and the catapult did not fire. *Gloria* spent several seconds immobile on the catapult before building up enough thrust for three seconds of slow ascent that ended with a belly flop. After the crowd waited another half hour for a refueling, a second attempt ensued. Although the engine performed better and the catapult worked, the plane "slid clumsily into the air," before somersaulting to the ground, well short of the state line. Kessler postponed a third attempt.

A reporter described the scene: "After this four-hour wait in a bitter wind," the disappointed crowd dispersed, offering "grumbles and disparaging remarks." The same reporter went to a nearby bar, where he talked with a former officer of the British Royal Air Force. This military-minded man mused about the application of rocket technology in a future war. Ley did not share his views about the military potential of *Gloria*. He avoided reporters. The stunt had been an embarrassing failure.[9]

Ley made a third attempt on February 23. Far fewer spectators and reporters braved the cold to witness the event. Nevertheless, Ley and Kessler staked a claim of grand success, after one rocket plane rose sharply before crashing, while a second plane simply slid over the ice, before briefly becoming airborne and then crashing at a distance of approximately 800 feet. The *Chicago Daily Tribune* reported: "More than 6,000 pieces of mail were carried across the New York–New Jersey state line today by two rocket propelled airplanes making short flights claimed as the first of their kind. . . . Fred Kessler termed the experiment 'successful.'" With regard to his motor, Ley agreed: "As far as the rocket motors were concerned, they have to be regarded as successful."

For "the story of rocketry," the Kessler flight was a day of "special importance." In Ley's words, it outshone the "flights of Dr. R. H. Goddard in this country and by Johannes Winkler in Germany." In his later memoirs Ley omitted these stunts from his histories of rockets.[10]

Ley discontinued his association with Kessler by spring 1936. In July he informed Pendray of his difficulties in finding other work as an engineer. According to Ley, the Kessler stunts resulted in a net loss, although they helped him make connections. Otherwise, Ley barely made enough money to cover the cost of living. He may have been working as a janitor or maintenance man, according to daughter Sandra Ley. Perhaps the manual labor reduced or covered his monthly rent. From March 1936 Ley occupied a small room in an infamous boarding house with a history of murders, fires, and strange events. The location was four blocks west from the south end of Madison Avenue, which includes Madison Square and the famous Flatiron building.[11]

Ley briefly considered a new scheme with public stunts. In October 1936 *Popular Aviation* reported on Ley's alleged progress with altitude rockets: "A series of rocket altitude shots—the first of their kind in this country—is planned by Mr. Willy Ley, associate of the famed Professor Oberth, pioneer rocket theorist." According to the article, Ley was further modifying the German "Repulsor" rockets. Ley planned to make a first attempt to reach 10,500 feet. The magazine reported, "The promise of complete success is very great."[12]

This article also claimed that Ley had just finished writing his first book in English, called *The Attack on the Stratosphere*. The book identified the next step in progress: meteorological rockets that collected data about the upper atmosphere. Then mail rockets "would be used to deliver mail across the continents or oceans in less than an hour." Next came the space rocket, "far in the future, perhaps, but sure to come, bearing with it staggering possibilities." "The weather rocket," Ley backtracked, "is, however, an immediate possibility . . . of not more than a year's work and of less an expenditure of money than the price of a large passenger airplane."[13]

The Attack on the Stratosphere did not find a publisher, and very little is known about Ley's continued experimentation with rockets in the 1930s. It is unclear how he might have obtained a work space. Perhaps his designs existed only on paper, and he hoped that the magazine article would generate inquiries from patrons who recognized his merits. Nobody came forth. Ley's further attempts to promote rocket experimentation met with skeptical audiences. For example, in a later article for *Startling Stories*, Ley recounted a hostile audience of engineers in 1936. The audience members made outlandish statements, such as: "All this is nonsense anyway. A rocket can't work in empty space. What has it got to kick against?" Ley expressed his astonishment at the unimaginativeness of American engineers.[14]

Meanwhile, Ley experienced a personal crisis as his visitor's visa expired. He could obtain neither a second renewal nor a permanent visa. Ley feared the possibility of being sent back to Germany. Perhaps he received advice from the Committee for German Refugees. Or he may have learned of a solution from other immigrants. They resolved the problem by leaving the country and reentering as Cuban refugees. The immigration loophole was risky and expensive. Ley had few resources. In a desperate attempt to shift gears and earn money, he wrote a science fiction story and several nonfiction articles for *Astounding Stories*, one of the more respectable pulps. Ley had always been a fan of these publications. He had been writing letters to *Wonder Stories* since 1930. He also appreciated traditions that began with Hugo Gernsback's desire to "promote a participatory vision of democratic science," as described by scholar John Cheng. When Ley could afford to buy the pulps for 15–25 cents, he did. Yet prior to this moment of personal crisis, he had kept a distance, still hoping to prove himself as an engineer. Between 1935 and late 1936 he had written several articles for more respectable publications, such as *Esquire-Coronet* and *Natural History Magazine*. Those publications paid well, but their acceptance rates were low. Now Ley was willing to associate his name with the pulps, while writing science fiction under the pen name of Robert Willey. He always claimed to have used a pen name simply to distinguish his fiction from nonfiction, rather than hide his identity. This claim is plausible, because editors did not disguise his identity.[15]

When Ley wrote his story and other articles, he desperately needed money. He feared being deported. He was still a German citizen. Talk of war with Germany filled newspaper headlines. It was a profound moment of personal and professional crisis. He needed to prove his loyalties.

At the Perihelion, Nearest the Scorching Sun

Ley wrote "At the Perihelion," a political and futuristic fantasy set on Mars. The tale recounts the adventures of American Dan Benson, who is stuck in the Soviet zone of the planet. Facing a looming deadline, Benson struggles to fill out an application of "Retainment of Terrestrial Citizenship." The application is described as "a cross between an American income-tax return, a German *Fragebogen* as to Aryan or non-Aryan ancestry and a G.P.U questionnaire for prospective members of the Russian communist party." The story adds: "It was a light year of red tape."[16]

By trade, Dan was a science writer exploring Mars: "It was about a thousand days that he had lived like a hermit in the Martian desert. One thousand days of hunting knowledge, hunting treasures . . . and pounding the typewriter." At this point, "he caught himself wishing to meet a girl." During a briefing with corrupt officials, Dan met a young woman, described as "a beauty. Just a beautiful Russian girl." Before Dan could speak to her, the Soviet officer announced:

> You are American, thirty-six, studied astronomy in America and have a German doctor's degree in chemistry. Your profession is that of a writer on science matters. You were a professor of astronomy at Columbia University and planned to marry about three years ago. Suddenly you resigned from your post, did not marry and went to Mars. Since then you were a fairly successful gem digger and a successful author.[17]

The Soviet adds, "Occasionally you write stories under the pen name of Herbert H. Harr."[18]

To avoid becoming a Soviet subject, Dan works as a contractor. The job involves exterminating the blue "skolopenders," which are deadly reptilian-like creatures. The young woman, Miss Nadya Tcherskaya, is sent with him as his supervisor. As Dan learns more about her, he grows very intrigued. She is not a typical Russian woman. "Where did you learn English?" he asked. "Fifty-Seventh and Broadway, mister," she answered, "I lived there for three years. . . . My chauffeur was a Negro from Florida, which added flavor." Benson studied her more closely, while falling in love: "She was so beautiful; there was no doubt about her intelligence. Suddenly, he stopped the drift of his own thoughts. He must stop thinking about her; he must never again start dreaming about her; she was—the most undesirable girl to fall in love with." As an intelligence operative, her mysteriousness and political savvy increase as the story progresses. She is smart, powerful, and exotic.[19]

The remainder of tale narrates their adventures while exterminating the blue menace. Arguably, the bizarre mixture of scenes reflected Ley's state of anxiety. On the one hand, it is an adventurous love story, as Dan and Nadya explore the ruins of an ancient Martian civilization. They fall in love as fellow explorers, intrigued by wonders of a dying world. There are moments of romantic comedy, as the two characters clash and argue. On the other hand, the story narrates a hellish world of forced conscription, corrupt bureaucracy, and state terror. For example, when Benson reports for duty, an officer asks, "Comrades . . . this is war. An enemy threatens the prosperity of Soviet territory. That the enemy is not human does not matter. It is war. But you may tell me who is too ill for duty." Those who step forward are shot. Others continue to live as slaves. Benson wonders when the slaves will turn their flamethrowers against their own tanks. Their world is a dystopian nightmare.[20]

Before long Benson is convicted of treason for falling behind schedule. He awaits his execution. Luckily, Nadya saves him, as the Soviet zone on Mars descends into open rebellion and chaos. Workers revolt. Armies fight each other. Dan and Nadya escape the Soviet zone via rocket ship. They first try to land in non-Soviet zones, yet new laws forbid the spread of anarchy outside the Soviet zone. Their only chance involves a trip back to the earth, on a path that takes them incredibly close to the sun. The last seven pages describe this harrowing space adventure of survival in the extreme environment of a spaceship. When they finally reach earth, Dan and Nadya marry in the United States. They live happily ever after. The Soviet nightmare is over.

"Scientifacts" and the "Scientifiction" Pulps

The pulps rescued Ley. Smith & Street publishers paid $235 for "At the Perihelion," along with another $185 for three factual articles. If adjusted for inflation, that sum is the equivalent today of $6,900. It is unknown if Editor F. Orlin Tremaine knew of Ley's troubles. He may simply have offered increased compensation to attract Ley. Regardless, the payment was generous. Ley left for Havana, Cuba (probably in early January 1937). Virtually nothing is known about Ley's time in Cuba, other than that he returned through Miami, Florida, on February 3. He disembarked from the vessel *Florida*. A clerk granted him lawful permission to enter the United States. When he arrived back in New York City, the situation became more stable. He could remain for the time being. He now knew his calling as a freelance writer. The pulps were his lifeline. For the next decade and a half Ley published numerous articles in *Astounding, Amazing Stories, Thrilling Wonder Stories*, and other magazines, before becoming exclusively contracted with *Galaxy Science Fiction* in 1952. Ley spent the rest of his life contributing to every edition of *Galaxy*.[21]

Some of Ley's earliest contributions to the pulps related to his recognized expertise on rockets and space travel. For example, *Astounding Stories* published a short nonfiction article named "The Dawn of the Conquest of Space." Soon *Thrilling Wonder Stories* proclaimed Ley as the "World's Foremost Authority" on rockets. A later article for *Astounding Science Fiction* also speculated on a "space war." Other articles included "Visitors from the Void," "Stations in Space," and "Calling All Martians!" For *Amazing Stories*, Ley wrote a seven-part exploration of the solar system.[22]

Incidentally, Ley accepted the title of "World's Foremost Authority" on rockets and space travel at precisely the moment when G. Edward Pendray wrote for magazines, such as *Sky* and *Scientific American*. Whereas Pendray's article "Number One Rocket Man" glorified the contributions of Goddard, Ley's "Eight Days in the Story of Rocketry" only briefly mentioned Goddard. In Ley's estimation Goddard did not deserve his own day in the history of rockets. Instead, Ley highlighted the contributions of Oberth, Valier, and himself. Pendray likely viewed his competition with Ley as a struggle to preserve the legacy of an American inventor. Eventually this competition hardened into a nasty feud.[23]

It may be tempting to focus on Ley's space-related articles in the pulps. After all, his sole claim to fame lay in his earlier association with the VfR. Yet it is important to note that the majority of his freelance writing between 1937 and 1941 displayed a much broader interest in the histories of science, technology, and exploration. Ley continued to branch out. He was becoming a professional science writer and a self-proclaimed "romantic naturalist." This branching out soon exploded in the science fiction pulps. Examples included a history of geology and continental drift called "Geography for Time Travelers," a history of "Earth's screwy plants!" titled "Botanical Invasion," an analysis of massive engineering projects labeled "Atlantropa—The Improved Continent," and even a speculative piece about "The Kitchen of the Future." It was also common for Ley to plead for the "The Conquest of the Deep," in several factual and historical articles about oceanic exploration.[24]

Additionally, Ley continued to publish articles for natural history magazines, such as *Frontiers*, *Fauna*, and *Natural History*. Examples included "Legend of the Unicorn," "Zoology of Wonderland," and "First Mention of the Giant Squid." Ley had several goals while writing these historical articles. Foremost, his income was stabilizing. Whereas the science fiction pulps typically paid about $10–$60 per article for non-fiction, magazines like *Esquire* paid as much as $112.50. On a more idealistic level, he wanted to encourage readers to share in his love of exploration. In a guest editorial for a science fiction pulp, Ley pleaded, "See Earth First!" He continued, "It is one of those pet beliefs of very many people nowadays that there is nothing left to explore or to discover—excepting, of course, discoveries that can be made in physical and chemical laboratories and those that are in the realm of astronomy." For Ley, it was simply bizarre that so many educated individuals believed that "on Earth the job is done." Ley continued:

Those that hold and voice this belief learn with almost a shock—and plenty of incredulity—that the surface of the Moon (meaning the four sevenths of it that we can see) is better known than the surface of our world. . . . The why is obvious, we see the surface of our satellite from quite a distance and we therefore see it always as a map while even stratosphere balloons do not possess enough elevation to see large portions of the Earth the same way. . . . We will have to wait for rocket ships.[25]

Nearly "one fifth of the land surface of the Earth is still unexplored," Ley wrote, adding, "On every continent—excepting only Europe—there are vast stretches of land that are either completely unknown or have been traversed only by the weary, worn-out, and fever-stricken explorers." Ley brought science fiction readers back down to earth: "We are now eagerly reaching out for the planets, at present in science fiction and in theory, a few decades hence in actuality. But in the meantime, before we are all ready to go out to discover other worlds, we have a job waiting: to finish discovery at home."[26]

In his own unique way, Ley participated as a rebel outside the ranks of institutional science. He celebrated the borderlines, the unexplored frontiers, and the great unknowns. He also promoted an interdisciplinary sensibility that contested what he viewed as a stubborn and unimaginative status quo of scientific skepticism. Natural history and exploration served as fruitful alternatives to classical physics and laboratory practices. Overall, he promoted the explorer as a scientific hero who embarked upon a journey.[27]

A Romantic Disaster

After Ley returned from Cuba, he married his first wife, a German immigrant named Margot Hübner. Little is known about the relationship and its brief union. By 1940 Willy and Margot were divorced. In a 1938 questionnaire for the Reich Association of German Writers, Willy described Margot as an Aryan Protestant, who worked as a font adjustor. He listed her political affiliation as RDP, meaning the Radical Democratic Party, a leftist coalition; its poor parliamentary performance had facilitated Nazi party victories in 1931. If Willy disguised her race, he could have disguised her political affiliation better.

Other public documents confirm that Margot was one year younger than Ley. She made a similar Cuban trip in 1938. Officials marked her complexion as DK, meaning dark. Exactly what that meant is unclear. She had brown eyes and brown hair. Otherwise we know very little about her and the failed marriage. There are strange stories that cannot be verified with documents. Other perceptions of Margot seem to imply a volatile German woman who espoused right-wing politics and hated New York City. Writer L. Sprague de Camp told the most bizarre story about Margot being pregnant with another man's child, forcing Willy to deliver the child in front of the real father.[28]

We may never know the true details of Ley's first marriage. He did not write about it in later years. It was a brief disaster, meant to be forgotten. He would go on to meet a different and exotic woman, whom he loved and cherished for the rest of his life. He called her a Russian beauty. They would raise a family together, through good times and bad.

The Search for Zero

Not only did Ley use the pulps as a financial lifeline during a troubled marriage, but he also used them to promote a broader agenda. He glorified unexplored frontiers by celebrating the great scientists of the past. Although these articles make today's historians cringe, they reveal the broader agenda of the time. For example, in a two-part article called "The Search for Zero," Ley summarized both science and its history in the following way. He began by describing a small book in his library. It contained "tables and figures and formulas, along with logarithms, measurements, and calculations." He concluded: "And that, gentlemen, is science." Logical inference could be made from the collection of data. The data demonstrated that "the world is an orderly world." A long but necessary period of collecting and recording explained why science developed so slowly. He then presented the perspective of other historians: "in former times, 'pure science' simply did not exist," because "what we now call the beginnings of science was in the hands of artisans and was, therefore, applied science." Science did not emerge until naturally curious individuals looked for the most perfect solutions "after realizing that the facts had to come first and the 'system' had to be molded to the facts."[29]

Science first progressed in the realm of astronomy, where the movements of the heavens obeyed rigid rules, confirmed by dedicated observation. Upon the discovery of an orderly universe, astronomy became the first science. Yet it remained the only real science for centuries, "because people had not then realized that other things and events also obey rigid rules." In short, "people lacked the conception of orderliness." The first step of any scientific endeavor began with a "basic conception of order." This "search for zero" was the search for a set of universal rules, "the rigidity of sequence," and the "starting point."

The path toward orderliness could contain pitfalls. Astrology, for example, projected the orderliness of the heavens on terrestrial events. Unfortunately, astrology "went haywire and did so with dire consequences to astronomy." In Ley's perspective, it degenerated back into the realm of philosophy. Astronomers had to struggle to correct its course. For the bold scientists, such as the brave Galileo, the Church was not the enemy. Astrologers and philosophers were their real opponents. Against a stubborn mental attitude of philosophers, the great men of science boldly offered the "destruction of mental security and the terrible realization that the road to knowledge was truly endless." To the chagrin of philosophers, the astronomers countered ideas with facts. They debunked false dogma.[30]

For chemistry, this struggle was harder. It had to "travel first all the way along the wearisome and disappointing road of alchemy . . . [which] began as fake and ended as one." The alchemists' endeavor "was plain counterfeiting, without the slightest shadow of self-deception." The chief sin involved secret facts communicated in secret language. Nothing served better to hinder scientific progress than a lack of openness and cooperation. Consequently, chemistry had the longest road to travel, "for what was regarded as knowledge was ballast in reality—ballast that had to be discarded as quickly and as completely as possible." For other sciences this battle for the formulae happened faster. Yet the struggles were not easy. They involved wars against human prejudices, mystifying language, and secretive practices. The scientist had to "find a way . . . out of chaos!" A science had to "thread its way out of that jungle . . . step by step." Most important, the explorers had to doubt everything, especially established knowledge. They collected anomalies in anticipation of a conflict.[31]

For Ley, the history of science did not consist of a smooth and gradual accumulation of facts, observations, or discoveries. The history of science documented conceptual shifts, caused by the accumulation of facts and observations. The real challenge in the ascent of humankind involved a conceptual shift. That new and revolutionary conception is the beginning (or turning point) of every science. Overall, science was a revolutionary and anti-authoritarian process, in which "many of the spiritual fathers of the revolution went rigorously on strike," refusing to show reverence for the great masters and philosophers. The bold thinkers spoke truth to power. There would be future revolutions in science. "There never is such a thing as an end" to the engine that drives scientific progress. So long as the freedom of inquiry flourished and authoritarian dogma receded, scientists could find the correct stepping-stones. They could destroy a rotten and false system of beliefs. They could build a new system on the garbage heaps of the past.[32]

The World of Tomorrow-land

Ley saw many visions of the future come alive at the 1939 World's Fair in New York City. This massive exhibition of the world of tomorrow impressed him greatly. He reported on the anticipation in a British publication. He wrote, "Only four years ago the several square miles of area that constitute the World's Fair grounds looked anything but interesting." "Flushing Meadows" was the garbage swamp of New York City. Marvelously, engineers had transformed the area into a site suitable for the display of scientific and technological wonders. Ley spoke optimistically about the "World of To-morrow," with its "future city." He wrote fondly of the "automatic machinery" that represented "a kind of robot civilization." He also appreciated the showmanship of the fair's optical illusions, spectator rides, and colored floodlights.

He expressed his hopes surrounding the Westinghouse Time Capsule, filled with books, photographs, and microfilm. It also contained a letter addressed to "humanity of the year A.D. 6939." Ley wondered about the reaction of human beings in 5000 years: "And it is quite possible that they may be impressed by a message from what they may consider a barbarian age."[33]

It is easy to imagine Ley touring the World's Fair with a sense of awe and wonder. As other scholars have noted, the World's Fair glorified Western progress. Exhibits like "Futurama" and "Democracity" offered visitors opportunities to tour the cities of the future from virtual heights, as they looked down upon models from balconies. Futurama was particularly popular because it was free. The ambitious design relied on grand spectacles and virtual witnessing to convince an audience to trust capitalism and modern science. These exhibits celebrated American modernism, consumer culture, and mass consumption. Corporations evangelized fairgoers with examples of modernist architecture, wondrous gadgets, and urban planning.[34]

These sites offered an experience of enchantment, wonder, awe, and appreciation of the technological sublime. Many of those experiences happened through virtual worlds that served as the playgrounds of the imagination. Quite often the technologies of flight took center stage. Not only did flight represent a pinnacle of contemporary accomplishments, but it also represented a collective experience of fairgoers, who peered down from the heavens to view the world of tomorrow. It was a transcendental experience. Visitors could witness a simulated dream world, in which anything seemed possible. In the words of scholar David Nye, the fair produced "a quasi-religious experience of escape into an ideal future equally accessible to all."[35]

Ley probably went directly to the Science and Education building, where he witnessed, in the language of a visitor's guide, "science . . . as a social force; as the new dynamic force which has chiefly created the modern world in which we live." The Education Exhibit announced: "Education in a democracy must be available to all men. It must train the whole man." The tour guide emphasized the relationship between scientific education and democracy. The tour book also commented on the timely need for a fair. It quoted a dedication speech that asked:

"How can mankind work and live in peace and harmony? How can life be made more secure, more comfortable, more significant for the average man and woman?" It answered, "This Fair, *your* Fair, is determined to exert a social force and to launch a needed message."[36]

These passages reflect the tensions of 1939. As scholars have noted, the unbridled optimism about the future coexisted with a fearful awareness of the present and the rising tide of war. Some 13,000 people per day toured Futurama. Many of them stayed close to a radio. "Democracity" was the engineers' dream. In Europe that dream was fading.[37]

The Fog of the Present

By 1940 Ley increasingly viewed Nazi Germany and the Soviet Union as two variants of totalitarianism. These perceptions can be inferred from a science fiction story called "Fog," published in 1940. Although the tale depicts a failed communist revolution in New York City in the late 1940s, Ley based the story on his experience in Nazi Germany. The conflation of communism and fascism was acknowledged by Editor John Campbell: "Ley . . . knows from first-hand experience the churning uncertainty of revolution's fog." The tale depicted "what a revolution in a major nation is really like."[38]

The story begins with a long quotation from a historian, who narrates the events of the Second World War. In this synopsis the United States remained neutral, while the rest of the world engaged in battles that spanned the globe. Hostilities then ceased, "due to complete exhaustion of all belligerent powers." The global situation led to a severe economic depression in the United States. Ideological fanatics took advantage of the situation.[39]

Readers are introduced to the main character of "the manager," as he concludes a disturbing phone call with Central Office. While his workplace is abuzz with chatter and gossip about the meaning of the phone call, the manager goes to lunch. Sitting alone at a restaurant, he overhears a heated debate in which a New Yorker says, "I have the right of free speech." A communist responds, "That's one of those contemptible bourgeois prejudices you cannot forget." He adds, "Free speech in political questions should be reserved for those with political schooling." The communist then argues that an untrained American has no right to debate with the architect of a new and ultra-modern building called Clemens Tower. The reaction of the manager follows: "He wondered—as he had done occasionally in business meetings— why Nature had not provided some mechanism to close one's ears as one could close one's eyes."[40]

The manager then returns to his office, where a colleague is waiting to discuss business. Yet before turning to important matters, they briefly discuss a young worker who distributes copies of *The Worker*. "He is good boy," the manager remarks, "but sometimes it looks hopeless." A character adds: "I wish times would improve quickly; steady jobs, with decent salaries, are to radical germs of politics what quinine is to malaria germs."[41]

When the workday ends, the manager walks home, thinking casually about the day's events as well as his wife and his pregnant sister. "Nature," he muses, "rarely suffered from depression . . . and always found a balance of some kind." The streets, however, show signs of imbalance. The police act nervously, "standing there in a fairly silent and entirely normal street, listening—for what?" Something is happening. The manager arrives at home and eases his mind by listening to his favorite radio program, an hour of soothing music uninterrupted by talk.[42]

The next morning, the manager grows even more confused. Both his doorman and his favorite newsstand vendor are inexplicably absent. The radio news seems entirely obsessed with a minor building fire. Most strangely, the gates to subway entrances are closed, and there are few cars on the street. After finally arriving at his office district, the manager has to convince a skeptical policeman that he is a legitimate employee of a firm at a verifiable address. He has to produce his papers. "What *is* going on here?" the manager asks. The policeman responds, "Sorry. Just orders. . . . Please move along." The manager tries to guess which building the police are protecting. Apart from his office building, key sites included the Radio Corporation, the *Daily Post*, the Union Building, and the post office.[43]

Finally, the manager arrives at the office, where he finds many nervous employees. He discovers that all long-distance calls have been suspended, while the radios have gone silent. Someone, possibly the police, has constructed a "Jenkins Radio Dome," which is a "propaganda antidote extraordinary." It serves as an "electric field inside of which all wave lengths beyond the red end of the spectrum ceased to operate." Suddenly comes the crackling of gunfire and explosions. On the city streets below, violence erupts. Then, just as quickly as it began, the violence ceases. Streams of employees exit the buildings to head home. The manager describes the scene: "Everybody was just hurrying home; not wasting a single breath on a useless word. . . . The only remark he heard on the way was a young man saying rather cheerfully: 'It ain't going to rain!' No, it wasn't; it never could rain in the area of a Jenkins Radio Dome." When the manager arrives home, he takes stock of the revolution. Phones and electric power are out. Store shelves are depleted. The radio dome casts a reddish hue upon the city. Meanwhile, trucks with searchlights roam the streets, determined to expose suspicious people. The manager can see a few prisoners being marched by the police. "To prison? Or to execution?" the manager asks. After hours of sitting in the dark, the manager falls asleep.[44]

He spends most of the next day trying to wait out the violence. But being desperate for supplies and especially cigars, he braves the city streets. Although he can hear distant gunfire, his own street seems quiet. Suddenly, bullets whine past amid shouts. The manager lunges into a doorway that gives way to Mr. Segal's cigar shop. Mr. Segal sits behind the counter and greets the manager. The manager asks, "Are you not afraid of the shooting?" Segal replies: "I am, I am, but what good does it? A bullet goes *zimm* through a closed door like though an open door. If it hits, it's God's will. What can I do about the bullets? It shoots here, it shoots there." The manager finds this advice comforting. He buys an extra supply of cigars and returns home. After a long and trying day, the manager closes his eyes. Only occasionally is he disturbed by the sound of random gunfire.[45]

A strange sense of normalcy returns the next day. Phone service and electric power resume. Gunfire ceases. And Joe, the doorman, resumes his duties, albeit dressed in a new costume: a red necktie and a red armlet. He greets the manager with a set of memorized slogans.

"Good morning, citizen," he calls out formally, adding, "The people have won and the Change is made. Now good times are here for everybody who works." "I hope you are right," the manager responds. The uniformed doorman explains: "Oh undoubtedly sir . . . citizen, I mean. Now the government is in the hands of trained masters, not elected amateurs. There will be a Victory Parade on Red Square, starting at eleven. It is advisable to be punctual." The doorman also explains that the new people's government is assembled in Clemens Tower.[46]

The manager discovers other facts. The *Daily Post* had been renamed as *Red Flag*, and its first issue contains almost no reliable information. He also discovers after a rather frightening encounter with a group of uniformed henchmen that attendance at Red Square is mandatory. An armed soldier calls out: "Today is a revolutionary holiday. . . . Better go to Red Square. Reorganization begins tomorrow." The manager then offers the soldier a cigar. A superior officer yells: "What does the citizen want?" The manager is then forced to show his hands, while the superior comments, "Wears a ring. Bourgeois. May have to be liquidated later on. Move on."[47]

Disturbed by this encounter, the manager approaches Red Square, where he notices hundreds of posters stating the official proclamation for all workers, soldiers, and citizens. After declaring that United States will be incorporated into the USSR, it advises all workers to return to work. Failing to comply will be recognized as sabotage. The new state bans all political parties, societies, and social groups except those official organs of the state apparatus. The state forbids travel and confiscates all weapons. It imposes a strict nighttime curfew, punishable by death. Additionally, it informs all members of the non-working leisure class to remain at home, awaiting a census officer.

For the remainder of the tale, the manager experiences the everyday realities of a totalitarian state. City services remain sporadic and unreliable. Stores remain sold out of bourgeois items. Propaganda replaces news and information. The only dependable aspects of the glorious revolution are the radio dome and constant renaming of sites. The manager still has a position with the firm, but he is now under the supervision of Sam Collins, the office boy who distributed *The Worker*. Sam's new managerial role involves making a daily speech aimed at boosting worker morale. The speeches get shorter and shorter. Other changes take place. The manager narrates, "People are arrested occasionally. Sometimes they come back and tell that it was all a mistake. . . . Or they don't come back. In the middle of the night you hear heavy boots on the staircase." During the next morning, "everybody insists that nothing happened at all during the night."[48]

The manager grows accustomed to these new realities. However, the glorious revolution soon falls apart. Although pockets of revolutionaries have established footholds in major American cities, the revolution simply cannot spread. Propaganda, in the United States, does not work. There is little popular enthusiasm to sustain the people's government. The American public rejects the ideology, while the United States Army does not side with revolutionary leaders. The Army soon retakes the cities in one quick sweep. Rational and anti-authoritarian attitudes prevail. A totalitarian revolution fails in a country that thinks rationally and scientifically. The story ends with an implicit celebration of the scientific minds of ordinary Americans. Americans think for themselves.

The Mysteries of the Past

In 1940 Ley wrote The Lungfish and the Unicorn: An Excursion into Romantic Zoology (1941). It would be the first book of a trilogy that he called "the adventures of a romantic naturalist." The book illustrated his true "romance" with nature during the 1930s. It also reflected his anti-totalitarianism, as he used the history of science as a weapon in a cultural struggle.[49]

Rather than focusing on mainstream zoology, the individual chapters "deal largely with the borderlines, with the vague boundaries of knowledge, with the twilight zones." While much of the book takes pleasure in debunking certain mythological creatures, other chapters are more optimistic about the discovery of sea serpents and living fossils. The most comical moments of the book describe bewildered scientists staring at strange fossils or wondrous specimens. Ley charted discovery after discovery of creatures thought to be mythological or extinct.[50]

To give his readers perspective, he outlined the history of zoology in a concise and entertaining way. The roots of zoological thinking began with tribal distinctions between good and bad animals. Then came pure human curiosity, when travelers told stories of strange and exotic beasts. Aristotle took the next step forward by cataloguing known facts and compiling lists, which represented the first stage of scientific thinking. However, "the first great landmark" came much later, with Pliny the Elder's *Natural History*. Ley clearly appreciated Pliny's debunking of Greek myths and legends: "The Roman Pliny had been a cavalry colonel and had wielded his stylus accordingly; if he had one firm belief, for instance, it was that all Greeks were liars." Pliny's debunking of myths and legends was a vital contribution to the science of zoology.[51]

Unfortunately, zoology then suffered through a "long arctic night from Rome to Renaissance." Ley's interpretation of the "Dark Ages" was clear:

> The Roman Empire collapsed. "Darkness fell," as the historians like to say, and the most valiant efforts of the Byzantines on the one hand and of the Arabs on the other could hardly preserve the knowledge gained, much less increase it and improve it. It needed the coming of the Renaissance . . . to bring a continuation of the "quest for new lands and strange beasts."[52]

The darkness was at last dispersed by the boldness of zoologist Conrad Gessner. Ley celebrates his contributions, arguing, "This man, whose broad mind was a mirror of classic knowledge, also started a new era." However, Gessner only contributed a list. The next stage did not occur until Carl Linnaeus "finally undertook the task" of naming all animals and plants, while defining their relationships to each other in terms of families, subclasses, classes, and orders. This eighteenth-century "achievement of simplification" helped to create "a unified system of knowledge." Linnaeus brought "rigidity and order out of chaos." He stepped in the right direction.[53]

All it took to move forward was a band of fearless evolutionists, who did not show deference to the founding fathers, including Linnaeus. Likewise, the later evolutionists lambasted French naturalist Georges

Cuvier for insisting on the stability of species over time. These evolutionists failed to appreciate how Cuvier himself had "crossed another and very important boundary line" by questioning if the earth may have had a very different past. In his "quest for new lands and strange beasts," Cuvier was "carried away by his own enthusiasm." Nevertheless, his system of closed-off periods of catastrophe and extinction led others to see the earth's past differently. Ley added: "Nonetheless we have Cuvier to thank for crossing the borderline between the animal world of today and that of the past, and for bringing to light the new worlds beyond it." The anomalies soon added up. Fortunately, it was not long before British geologist Charles Lyell "succeeded in . . . wrecking the whole theory beyond repair." While being influenced by greater estimates of the age of earth, as well as notions of extinction, he "quietly discarded the slightly hysterical outbursts from the Continent and replaced them with *time*."

For Ley these events showed a pattern in the history of science. On the one hand, science matured through anti-authoritarian explorers, who boldly questioned the wisdom and authority of experts. The enemies of progress resigned in closed-minded deference to figures like Aristotle and Cuvier. On the other hand, there was clearly a relationship between these great men, whose own contributions to science helped humanity to ascend, step by step, toward a more perfect and truer understanding of the world. "Under the scientific method Truth is an absolute thing," Ley later asserted. Experts always demarcate the boundaries of science, yet it took bold adventurers to cross the borderlines, ask new questions, and seek out the unknown. If these endeavors were done in rational ways, the borders of science expanded and conceptual revolutions ensued. If these quests were done in irrational ways, the adventurers were cast adrift in a sea of pseudoscience and cranks.[54]

The bulk of the book crossed other borderlines, with case studies in the history of science. From unicorns to famed sea serpents, Ley takes readers on a wild ride through the history of science and exploration. He debunks or defends eyewitness accounts. He tells entertaining stories about the discoveries of living fossils. Some of these strange creatures had survived even the harshest of geological catastrophes.

Others had recently become extinct. A central thread laces the book with a sense of wonder and mystery. The thrilling mysteries of science exist both in faraway places and all around us. Something entirely unexpected might be discovered. Monsters could wash ashore. Readers are left wondering: Who knows what might be lurking in the zoological shadows or the oceanic depths? What else is out there? What other types of dogma will be discredited? What conceptual revolutions will follow a new age of exploration?

The Days of Creation

Ley advanced this romance with nature further in his 1941 book, *The Days of Creation*. It presented an entertaining and holistic biography of our planet that compared geological periods of the earth with the Book of Genesis. A superficial scan of the table of contents would indicate that Ley was attempting to reconcile science and religion, with chapters called "Let There Be Light!" and "The Glory of the Mammals." Overall, the book makes the case that "these two accounts . . . are remarkably alike as far as the sequence of events is concerned." Thus the chapters discuss the birth of the sun and earth, the evolution of oceanic life, the conquest of the land by vegetation, and so on. It is a history of evolution that celebrates the similarities between modern science and ancient mythology. Ley did not advocate for the divine truth of scripture. No theologian, he argued, can seriously take the biblical account at face value. However, Ley went on to celebrate the remarkable fact that nineteenth-century science independently established a history of the planet that matched the days of creation, if days are equated with very long geological periods.[55]

Ley spoke directly to his readers: "Come out into the night with me, away from your reading lamp, and let us look at the stars." He continued:

> The night may be warm, but even then you will experience a slight shiver of . . . no, not of cold. It is something else, something for which I do not know a perfectly fitting word, neither in English nor in any other language. It is a sensation of infinity; a sensation which

does not exist in daytime. A sensation which has moved philosophers and inspired poets and which has, at least once, set everyone
thinking.[56]

Ley then presented the history of astronomy as a history of that sensation of infinity, as the realization of a plurality of worlds led to a realization of a plurality of galaxies. For Ley it was a story of the evolution of
man's consciousness. He humbly admired the extreme vastness and
the complexities of nature. The discoverers of that vastness and complexity also felt humbled by billions and billions of other worlds. Ley
inspires readers to feel likewise, as he takes us to the distant past of
geological epochs.

The themes of the chapters reinforced Ley's historical perspectives.
For example, the chapter "Let There Be Light!" is simultaneously a
history of the universe and a history of human enlightenment. The
history of astronomy involved a battleground of competing theories,
which "might be likened to a succession of encounters between battleships of increasing modern design." Likewise, the stories included the
planting of mines that led to conceptual revolutions. The gradual discovery of the age of the earth was just such a story, comparable to the
discovery of the universe. *Days of Creation* tells the fascinating story,
while inviting readers to imagine the alien landscapes of past geological periods. Readers are invited to marvel as the first plants and then
animals ventured out of the ocean. The later ascent of humanity is no
less wondrous than the prior triumph of the reptiles.[57]

Humans would continue to ascend and adapt. Readers should be
skeptical of doomsayers who claim otherwise. As the history of science
proves, "Man . . . is adapting himself rapidly and most efficiently with
new inventions all the time, and he remains the same, too, ready for
further adaptations." This journey of exploration, self-discovery, and
reverence for the complexities of nature would continue unabated.
Science and technology would serve as the technological means for
adaptation and exploration. Nature lay ready to be further tamed and
exploited. The explorers had a God-given destiny to ascend.[58]

The Broader Agenda

Ley wrote these books for a broad audience. Rarely did he openly admit an agenda, beyond entertaining and educating intelligent lay readers. Nevertheless, we can see that he popularized a very positive image of science, along with a celebration of the key figures who contributed to the ascent of human beings. In this regard he belonged to a larger camp of scientific intellectuals who shared a vision of science as a unified human endeavor that unmasked a single reality. They shared the goal of bringing science to the people, in an effort to transform and enrich culture. This camp included other émigrés, such as members of the well-documented "Vienna circle." Other scholars have described the ideals of émigrés and American-born intellectuals who campaigned for a more secular and scientific world. Quite often they advocated for the "scientific attitude" or the "scientific spirit" by celebrating reason, freethinking, and democratic participation in science.[59]

This generation of Europeans had perceived the spread of irrationalism, pseudoscience, and mysticism in Germany. They also witnessed the role of mass media in the spread of dangerous and hateful ideas. Now, after escaping that context, they witnessed other events, such as the alleged public hysteria during Orson Welles's famous radio adaptation of *War of the Worlds*. Although the actual extent of the panic can be debated, the newsprint coverage demonstrated (to many intellectuals) how an enormous deficit of scientific knowledge and critical reasoning accounted for mob behavior and irrational action. Elitist perceptions of mob irrationalism circulated widely among intellectual émigrés. They sought allies within the American intellectual community who also promoted scientific thinking. Instead of encountering resistance, they found much common ground with American intellectuals who wrote popular books and articles. They sympathized with the efforts of men like philosopher-reformist John Dewey, who embarked on a quest to make the world more scientific.[60]

Many Americans shared their perceptions of a totalitarian menace. As documented by other historians, fears of totalitarianism solidified prior to the Cold War, when it became common to lump communism and fascism together as the same political monster. Later cultural critics further associated totalitarianism with mass culture, broadcast media, and state propaganda. While these associations became particularly hegemonic during the Cold War, the roots of this political culture and ideology predate the Cold War. Ley fits well into a broader camp of cultural producers who influenced a broader public. His political identity was shaped in opposition to totalitarianism and pseudoscience. Over the next decade he communicated his anti-authoritarian beliefs to a mass audience, hoping to counter nonsense or hysteria. He continued to associate communism and fascism with the spread of irrationalism and pseudoscience. Yet Ley never blamed popular media or the inherently passive audience. Instead, he embraced mass media, while never losing faith that the vast majority of people could make informed, rational, and fundamentally scientific decisions, if they had been trained to do so. They needed to think scientifically about the world. They needed an expert who helped them to think for themselves. The science writer could fulfill this role, not as a dictator of facts but as a teacher who inspired critical thinking and independent thought.[61]

Ley's perspectives reveal how popular representations of science and scientists were also shaped in opposition to a totalitarian menace. The key distinction between totalitarian crowd psychology and rational democracy rested in the public's embrace of scientific thinking. Communicating the scientific spirit in popular realms became a tactic to preserve democracy and science from the perceived opposites of totalitarianism and pseudoscience. Arguably, Ley's shift from a specialized subject to a broader popular front was not a unique career change for many scientific intellectuals. He shared many of the goals of this camp. He campaigned for a more scientific world. He appreciated the opportunities that mass media provided. He spoke directly to the public, attempting to promote a version of scientific thinking that was neither stripped of wonder nor confined to a journal, institution, or field of study. In the pages of science fiction pulps, general and natural history magazines, and books aimed at both a juvenile and adult public, Ley explored the history of science, while promoting his own versions of the ideas that circulated in discourse. He celebrated the unity of humankind, as embodied by science. He also celebrated the "wholeness" of science. Most of all, Ley promoted a code of the scientist that meshed well with the later views of social scientist Mark A. May, who wrote, "Let all mankind imitate the fellowship of science." Yet, to the dismay of Ley and others, humankind did not imitate the fellowship of science. Instead, madness spread. The world went to war.[62]

The *PM* Years and the Science Writers at War

In May 1940 Willy Ley joined the staff of the political tabloid *PM*, a leftist and provocative newspaper. The publication was the brainchild of editor and social activist Ralph Ingersoll, who believed that advertising undermined the objectivity of the press. Instead of relying on capitalistic power brokers, *PM* would depend on its readers by charging five cents for daily editions and ten cents for special issues. Accordingly, the tabloid claimed to present the unvarnished truth, unsullied by war profiteers or media conglomerates. It also claimed to represent the last vestiges of serious journalism, combined with a rational evaluation of facts.

The tabloid often stated its manifesto:

> This Is *PM*.... We are against people who push other people around, whether they flourish in this country or abroad. . . . We do not believe all mankind's problems are now being solved successfully by any existing social order, certainly not our own, and we propose to crusade for those who seek constructively to improve the way men live together. . . . We are Americans and we prefer democracy to any other principle of government.[1]

Despite its denial of party affiliation, the publication had a clear and consistent crusade during its early years. In fact, the editors of *PM* could easily fit within a broader camp of "voluntary propagandists." Foremost, *PM* sought to expose the war crimes of Germany, while pleading for U.S. entry into the European conflict. *PM* screamed, "35,953 Innocent Men, Women and Children Killed by Fascist Bombs . . . What Are We Going to Do about It?" To shock readers, *PM* printed full-page pictures of dead women and children.[2]

Not only did *PM* denounce fascism abroad, but it also exposed fascist threats at home. Headlines advocated for immediate action against a right-wing "Fascist Front" in the United States. Favorite targets included aviator Charles Lindbergh and radio-priest Charles Coughlin. Anyone who openly sympathized with the Nazis or voiced isolationist sentiment was branded as a domestic enemy. From *PM*'s perspective, the sympathizers openly plotted a revolution while blanketing the country with propaganda though Hearst's monopoly of newspapers and magazines. In the face of these threats, *PM* aimed to educate the American public, debunk propaganda, and generate support for intervention.[3]

Predictably, *PM* attracted a large cast of idealistic journalists, including outspoken writers I. F. Stone and James Thurber, along with illustrator Theodor Geisel (better known as "Dr. Seuss"). *PM* also supported a large cast of photographers, sports experts, and media critics, whose work filled the pages of weekend editions, when the tabloid took a break from scandals and political rants. Its longer Sunday issues printed more pleasant articles as well as large photographs, including scantily clothed pin-up girls. Readers could also receive updates on *PM*'s adopted baby, Lois, as she developed from infant to toddler. Additional content included public interest stories and fashion advice.

In a letter to an editor Ley described his initial role: "I represent science and aviation in the radio-advertising Presearch [*sic*] Department which is . . . really the Department of the Future. We are concerned with things to come." Although he would later become *PM*'s "science editor," he joined the staff as a researcher. He informed Robert Heinlein, "It is, believe me, woefully hard to convince newspaper people that they should open some space for science . . . with a war and conventions going on and more war and a presidential campaign with broken traditions coming up—and I feel somewhat hrt [*sic*]."[4]

Ley spent a great deal of time at the offices of *PM*, where he worked "the night force," from 3:00 p.m. to 11:00 p.m. Much of this work involved fact checking and perhaps translation duties. While he campaigned behind the scenes for more scientific content, very few articles on science appeared prior to January 1942. For the editor, the war cause was simply too important to devote much daily space to non-essential content. Occasionally *PM*'s weekend edition might accommodate Ley's interests. Otherwise the tabloid had a solid formula of updating readers on the war front, while lambasting American publications that helped the enemy's morale.[5]

Historians have wondered why Ley associated himself with a leftist publication that employed communists while engaging in political slander, accusations of treason, and even a brief battle with the army for drafting its editor. Ley did not sympathize with the political leanings of some staff members, as can be seen in his science fiction. It is tempting to imagine that he studied some of the staff as if they were bizarre human specimens. When Heinlein asked him about an article that reported on a talking dog, Ley responded: "It's not a hoax." Ley described the story's reporter: "He's a communist and consequently devoid of any kind of imagination as I found out in many conversations. . . . If O'Connor says a dog talked in his presence that dog most decidedly did." Ley regarded communists as unimaginative followers of dogma.[6]

It is tempting to view Ley's relationship with *PM* as one of convenience. It was his first steady job in the United States. Being on *PM*'s staff may have allowed him to do much of the busy work for his books and articles. Additionally, it was very convenient for Ley to associate himself with one of the most anti-fascist newspapers when he was still an enemy alien, fearing the possibility of deportation or internment. If Germany declared war on the United States, Ley might be detained or imprisoned, as his father had been during the Great War. In some ways Ley's role for *PM* was self-serving, convenient, and even apolitical.

Nevertheless, a closer look at his evolving relationship with the tabloid reveals a more complex portrait of a science writer during the Second World War. If one label could describe Ley's public persona from late 1941 to early 1944, it would be that of a war weapons expert. He wrote far less about space travel and natural history. There are important exceptions to this trend. For the most part, Ley became immersed in technical journals, military histories, and news of the war front, as he attempted to educate Americans, alleviate public anxieties, and debunk German propaganda. Although his down-to-earth explanations of war technologies rarely contained overt political statements, his science writing cannot be divorced from politics and anti-authoritarian beliefs, particularly when compared to a broader scene of New York–based science writers. His central message of "Keep Calm!" also related to his firm belief in the power of scientific thinking, especially during times of fear. Ley was deeply fascinated by the technologies of war as well as the publishing world of New York City. He embraced the role of the science writer as an interpreter of information. Ley had a duty to enlist in the war effort through responsible journalism. In his view, he provided a vital public service.

"Presearch" and Publications

Throughout early 1941 Ley worked mostly as researcher, occasionally writing short columns for *PM*. He fact-checked articles and corrected misinformation. Then his direct writing increased briefly in summer 1941, when the tabloid introduced Ley's "New Weapons Department." The editor invited readers to submit their ideas for future war weapons "designed to lick the Nazis." Ley would judge the merits of the readers' concepts. The column almost always included an illustration of engineering components or trajectories. Incidentally, the first installment criticized a reader's suggestion of "rocket cannons" as too expensive. "What could it do that existing weapons cannot?" Ley asked. Planes could deliver a far greater amount of direct hits. In a different article for the *Coast Artillery Journal*, Ley reaffirmed his convictions about war rockets: "Rockets as a weapon of war are nowadays about as obsolete as catapults and crossbows." However, Ley admitted that "some form of rocket may be evolved to take the place of night bombardment aviation."[7]

In nearly all of his "New Weapons Department" articles, Ley gave a thumbs down to the ideas of readers, who seemed to misunderstand the mechanics of existing weapons. Readers presented unrealistic ideas, often based on misinformation from popular books. Ley corrected their mistakes, without condemning their lack of knowledge. For example, he patiently explained a reader's idea of "Bat Men for Parachute Invasions." According to this idea, a soldier could employ batlike wings to assist in gliding during a parachute jump. Ley wrote, "I doubt whether this impressive sounding weapon would have any real value under actual battle conditions. . . . Thumbs down on this one." This column was not popular, and it did not last long.[8]

What emerged from his behind-the-scenes "Presearch" on weapons can be read most directly in a book that Ley wrote in 1941: *Bombs and Bombing*. As a "brisk, popular survey," the book explained the various types of bombs being used in the war as well as the means of defense against them. Ley also examined the mechanisms of weapons, the physics of explosions, and the effectiveness of various types of shelters. Overall, it was a fairly dry and educational book. Instead of encouraging a sense of wonder about nature or modern marvels, Ley moved from one discussion of weapons to the next, attempting to educate readers. Only one section of the book brought Ley's personality to the surface, when he attempted to debunk and dethrone the popular prophets of doom and their "'horror' novels" that created hysteria about gas warfare. Ley aimed his sights most directly at H. G. Wells, whose *Shape of Things to Come* (1933) contained "a skillful symposium on chemical horrors." Ley stated, "I hasten to assert and emphasize that *none of Mr. Wells's statements contains even a grain of truth*, save for the one which says lewisite was discovered by Professor Lewis of Chicago and that it was not used in the first World War." He added: "But Wells's ridiculous nonsense reflects what many persons still believe about poison gas." The remainder of the book reverted to a dispassionate tone, as Ley explained the facts, while discrediting the military effectiveness of chemical weapons.[9]

The calm, objective, and scientific tone of the book caused one reviewer to call it "morbidly interesting and readable." Despite the lack of any sense of moral outrage about destruction and the loss of life, the book could be read as "reassuring and encouraging," because an air raid on New York City "could hardly be half as bad or as devastating as most people imagine." Indeed, Ley's main message in the book was put in bold-faced print in the second edition. "KEEP CALM!" he yelled. Reviewers found this message comforting. Not only would ground defenses be effective counter-measures to aerial bombardment, but also the effects of bombs and bombing were not completely devastating. A reviewer summarized: "The fact is . . . that the worst air attack imaginable could do only a certain amount of harm, unless it went totally unopposed." Another reviewer commended the book because it dispelled the "the vague terror which is founded on ignorance." A different reviewer went further: "This valuable little work should . . . calm the fears of the hysterically apprehensive." There was no reason for panic.[10]

A Brief Moment in Time and a Russian Beauty

In October 1941 Ley was a bachelor on vacation. Following a battle with the flu and a throat infection, Ley admitted, "I need California sunshine." He boarded a bus. After a brief pause in Chicago, he traveled to California to visit film director Fritz Lang, now living in Santa Monica. Although separated by a great distance, Ley and Lang remained close compatriots. According to Lang, they had hour-long conversations on his terrace, where Lang frequently pointed to the moon, calling it "my location set." Ley responded, "We will be there!" After staying with Lang for two weeks, Ley visited Heinlein in Hollywood. From their correspondence, it seems clear that Heinlein and his wife intended to introduce Ley to a writer named Virginia Purdue. Ley was searching for a female companion.[11]

It is possible that Purdue accompanied the Heinleins and Ley on a trip to "the forest of the giants." It was one of Ley's most memorable romantic excursions, in which he became a witness to the ancient past. He remembered, "We stopped for the night in Visalia and all through the night there was a howling storm with sleet and rain. I should have expected that from latitude, altitude, and season. But reasoning needed some time to overcome an established mental picture; for hours I was experiencing a kind of prolonged wondering surprise." When Ley spent much of the next day touring the sequoias, he felt a sense of "timeless vitality." Unlike ordinary trees, the sequoias did not "seem to cower under the clouds, waiting for them to disperse." Instead, "They reached up and supported them." Ley continued: "The gigantic columns of living wood, their green boughs partly obscured by the cloud veils, seemed to create a strangely roofed island. Not an island of mountain forest . . . but an island in time, an island in the time stream which flowed around them. And also an island of silence." Ley displayed reverence for nature, adding, "One does not speak loudly in the forest of the giants." Overcome with the gift of a "special permit," Ley spent "a few hours in the past." It was a quasi-religious experience.[12]

On November 3, 1941, Ley flew back to New York on a TWA "Stratoliner." It may have been his first time as an airplane passenger. He described the experience as beautiful and relaxing. At some point during the following weeks he met Olga Feldman, a young Russian woman with a background in ballet and dance. She had joined the staff of PM as a fitness columnist and model. While her pictorials offered useful advice, the photographs of her exercise poses could be provocative. She often wore a mask. Many of the photographs are indistinguishable from pin-ups of the time, apart from the written advice on exercise. Ley informed Heinlein: "I found her in our office when I came back from Hollywood. . . . She is Russian, born some 29 years ago in what was then still St. Petersburg." Ley further described Olga:

> She is in this country since 1920, citizen "by derivation" (since her parents took out papers while she was still a minor), is utterly reliable, lacks the proverbial Russian temper completely, flawlessly good-looking (I avoid stronger terms which may seem prejudiced), speaks English, French, German, and Russian flawlessly, modern Greek almost fluently and has an IQ in the neighborhood of 175. Are these reasons enough? As for me the most important one is that she loves me. And that she is old, experienced and intelligent enough to know what that word means when she uses it.[13]

They became engaged on December 11, the day Hitler declared war on the United States. They married two weeks later, on Christmas Eve.

Olga moved into Ley's small New York apartment, while both continued to work for *PM*.

A Science Writer Enlists

Although Ley was still an enemy alien, he was now married to an American citizen. Yet he still feared deportation or internment. On January 14, 1942, Roosevelt issued Presidential Proclamation 2537, which required all enemy aliens to report changes of address, employment, or names to the FBI. Soon Executive Order 9066 authorized the creation of exclusion zones, which led to the internment of both enemy aliens and American citizens of Japanese ancestry. There was much talk in the streets and press about a possible German internment. *New York Times* headlines indicated that the FBI was beginning to question and detain Germans and other European enemy aliens.

At precisely this time Ley began to write fairly regular and much longer articles for *PM*. It is unknown whether this increase of articles was due to his initiative or the goals of the editor. It was probably a combination of the editor's desire to educate readers about the realities of warfare and Ley's attempts to associate himself publicly with the most pro-war and anti-fascist newspaper in New York City. It is also noteworthy that Ley's efforts to educate the American public about the technicalities and dangers of war weapons came at precisely the point when other publications, particularly *Astronautics*, suspended their presses. President H. Franklin Pierce told members of the American Rocket Society: "Because of the military potentialities of rocket power it is deemed essential that the dissemination of further information on the subject be curtailed." He then urged his fellow enthusiasts to "use discretion in talking of past experiments, or in giving any information relative to rocketry which might be of aid to the enemy." While the ARS became silent, Ley offered readers a sober examination of the potential of war rockets as well as much information on the history of their designs, fuels, and uses. Most likely he viewed the ARS's silence as silly, considering that most of the material was readily available in public libraries and archives.[14]

In *PM* Ley also reiterated the themes of *Bombs and Bombing*. His "War Weapons" articles explained the technologies of war and the tactics of enemies. In these articles he doubted the effectiveness of most types of gas bombs as well as chemical warfare: "Could not the whole war be won by decimating (if not exterminating) the population of all the big cities by means of germs? The answer is NO." On the effects of gas bombs, in particular, Ley waged a public campaign to dispel nonsense that had been published in many books. He explained, "When the present war started, most people expected that things they had read in books and magazines for many years now would be terrible truth. They believed in all seriousness that the new war would be a 'poison gas war,' a war not only without victors, but a war without survivors." Public misinformation was the result of a "systematic campaign carried out by countless would-be prophets of the 20s." Chief among the offenders was Wells, who profited from public anxiety. Ley wrote: "Truth was a minor consideration in this campaign; sensationalism was what counted." Instead of offering fantastic nightmares of future wars to come, Ley countered with sobering facts and calming predictions.[15]

In the midst of Ley's attempts to debunk earlier works, a book became a bestseller: Major Alexander P. de Seversky's *Victory through Air Power* (1942). The title indicates the central theme of the book. When Ley read this book, he immediately sat down to write his own book-length response, *Shells and Shooting* (1942). In the introduction he explained his motivation: "Recently the writer of a highly overpromoted and forcefully circulated book asserted with great vehemence that the age of artillery is over now, that airpower—the bombing plane—has taken over and that heavy batteries, fortified positions, and anything afloat (especially battleships) are outclassed, obsolete, and a waste of money." Ley responded: "I cannot subscribe to such a thesis." Ley debunked Seversky's thesis by presenting a detailed survey of ground weapons that could counter aerial bombardment. On the subject of war rockets, Ley concluded: "It is not likely that rocket artillery will be revived during the present war."[16]

Ley presented a much more forceful attack on Seversky's thesis in the pages of *PM* on December 9, 1942. The four-page article,

"Debunking Seversky's 'Victory through Air Power,'" stated directly: "It is important that people get the facts straight." Ley classified the book with "50 such books" produced during the interwar years. "They all made exciting, if gruesome, reading," he wrote, "but as for actual value—military or prophetic—none was worth the paper it was printed on." He also accused publisher Simon and Schuster of a dishonest letter-writing scheme as well as an unbelievable advertising campaign. He argued: "The book saw a promotion as no other book in all history, with the possible exception of *Mein Kampf*. It was crammed down the throat of John Q. Public." Simon and Schuster profited well from a fear-mongering prophet of doom and his "old horror soup."

Ley then compared Seversky's book to the propaganda of Joseph Goebbels, while accusing Seversky of undermining the war effort: "It is clearly evident that Seversky has little if anything to contribute to the war effort. It is just as evident that his scathing . . . criticism of our High Command has done great harm to the *morale* of considerable numbers of people." In concluding his powerful debunking of the book, Ley illustrated an airplane of the future as imagined by Seversky. It included five heat rays, four secret weapons, two disintegrator ray projectors, a movie projector, six atomic motors, and a "pilot, just for emergencies, mechanical brain does all the thinking." A few days later *PM* published a one-page response from Simon and Schuster editor Quincy Howe. It contained a rather weak defense of the book, along with an attack on Ley for getting a few facts wrong and for comparing the book to *Mein Kampf*. Ley responded to the article by stating, "I am surprised the passing mention of *Mein Kampf* hurt so much. It is nevertheless a fact that young air power enthusiasts quote Seversky in precisely the same manner as ardent young Nazis quote Hitler."[17]

Following this incident Ley received an Order to Report for Induction in January 1943. Unlike his editor at *PM*, Ley displayed no moral qualms about joining the United States Army. He may even have considered it an opportunity both to apply his expertise and to demonstrate his loyalty to the United States. He was still an enemy alien. In a letter to Editor Frederik Pohl, Ley casually stated, "PS. I'm reporting for inducting Thursday. So Long . . . unless I am rejected because of poor eyesight." Pohl replied, "Best of luck to you in the Army, Mr. Ley, should you be accepted." After twice reporting to the induction center, Ley received a physical exam. In a different letter to Heinlein, Ley recalled: "As for the Army, I was in it for precisely four hours, the time it took them to put me in 4F because of poor eyesight."[18]

After this rejection Ley continued to write many more articles, most of which contained illustrations and diagrams that dissected grenades, mortars, mines, and guns. Other articles focused on potential weapons of the future. On rockets, in particular, Ley wrote: "Rockets have almost become a symbol of new weapons to be feared. Hitler's Danzig speech, which darkly hinted at 'secret weapons' of the Germans, was widely interpreted as referring to war rockets." In other articles for *PM* he simply educated the public about the physics of shrapnel, the limitations of germ warfare, and the effectiveness of searchlights. At every conceivable opportunity, Ley tried to debunk "Propaganda Weapons" of the Nazis. For example, he evaluated reports that Germans were testing poison gases on the Eastern Front. "Many observers," he explained, "believe that it was the beginning of a large scale use of gas. . . . But it is at least as likely that it was just a test application of poison gas, staged not for any direct military reasons but to the benefit of the analysts in the German Dept. of Psychological Warfare or the German Ministry of Propaganda." Ley focused his next eight articles on gases and the defense against gas bombs. He provided the public with basic steps on counter-measures, should such an ineffective weapon be used.[19]

Ley's writings made an impact on the publishing scene of New York City. The editors of *Mechanix Illustrated* labeled him a "War Weapons Expert" while stating, "These articles and his bestselling book of last fall, *Bombs and Bombing*, have been so widely hailed as definitive discussions . . . that MECHANIX ILLUSTRATED asked him to devise and describe the weapon which in his estimation could stop the tank terror." Ley used this opportunity to contribute regularly to the magazine, which published more of his articles on "war rockets" and "super guns." He also contributed articles to *Astounding Science Fiction*, including "Bombing Is a Fine Art," "The Paris Gun," and "Terry Bull's Terrible Weapon." Ley reassured readers by making them understand the scientific and technological facts about the current war. He dispelled hysteria, while exposing the war profiteers who frightened the general public. If his readers were presented with the contemporary and historical facts, then they would realize that there was no reason to panic. By 1943 Ley must have felt an enormous amount of pride for his public service. Germany was losing the war, and most Americans felt less apprehensive. All talk of "terror weapons" under development was sheer nonsense.[20]

The Spaceship of the Future

By 1944 Ley earned a consistent monthly income as a science writer. The year 1944 also marked several important and happy events in Ley's personal life. On March 11 Olga gave birth to a daughter named Sandra. Three days later Willy became an American citizen. He finally felt secure. Not only was he enjoying a successful career as a public educator, but he was also in the final writing stage of a memoir of sorts. This autobiographical book began as a two-part article for *Astounding Science Fiction*. In "The End of the Rocket Society" Ley recalled the rise and fall of the Verein für Raumschiffahrt. By spring 1944 this autobiographical account of the VfR had grown into a definitive history of rockets. Ley titled the first edition *Rockets: The Future of Travel Beyond the Stratosphere*. Although the first edition was not an instant hit, it would become one of Ley's most influential and popular books. Eventually this book expanded from 271 pages to 557 pages. Viking would release twenty-one editions. It also became an international seller, translated into nearly every European language.[21]

Rockets presented a definitive history of rocketry, from early theories of a plurality of worlds to recent engineering accomplishments. Ley wrote:

It is the story of a great dream, if you wish, which probably began many centuries ago on the islands off the coast of Greece. It has been dreamt again and again ever since, on meadows under a starry sky, behind the eyepieces of large telescopes in quiet observatories on top of a mountain in the Arizona desert or in the wooded hills near the European capitals. It has been dreamt all over the earth, in places ranging from quiet libraries to noisy machine shops. And everyone who thought about that dream added a little knowledge.[22]

Ley wrote: "It is also a story of continuous progress, one small step here and another one there." Not only was it a story of a dream; it was also the history of a technology that "evoked different ideas in the minds of people at different times." According to Ley, the rocket had long fluctuated between two extremes: "the grim weapon of war and the instrument of amusement." Yet this dual identity was suddenly unimportant. A new era of exploration had dawned. Although "there will be war rockets and amusement rockets in the future too . . . there will be bigger and more important applications than either of these two." Ley predicted, "And as for war rockets, in spite of some spectacular applications in the present war most of their story lies in the past." Ley concluded, "I'm going to speak about spaceships. Some time in the future they'll exist."[23]

What follows is an entertaining romp through the history of science. It is perhaps one of the most Eurocentric histories of astronomy. Whereas the Babylonian "conceptions were childish," Chinese astronomers "did not even guess that the lights in the sky which they observed so diligently might be other worlds," and it was up to Greeks "to invent better concepts, concepts which coincided to a large extent with reality and served as a starting point." While the Greeks "almost succeeded in arriving at a true picture of the solar system," their philosophical speculations complemented their astronomical observations. Unfortunately, scientific progress was severely hindered by medieval followers of Aristotle. Ley wrote:

It literally came to a point where thinkers set out with the notion that all wisdom could be found in the Bible, all astronomy in the Almagest, and all science in the writings of Aristotle. Not only was it simply forbidden to teach anything that contradicted or diverged from Aristotle's statements, it was also denied that there was anything that Aristotle had not known.[24]

Due to closed-minded adherence to a learned authority, science was at a standstill. It would remain stagnant until "the astronomical revolution" of Copernicus, who created "a new picture of the world," further developed by Galileo and Kepler. Ley then ruminates on Kepler's science-fantasy, *Somnium*, to illustrate "the new telescopic era which dawned." It was a new age of rediscovered texts and experimentation. It was also a new age of fantasy and science fiction. For Ley these two trends were deeply connected as Copernicanism spread throughout Protestant countries. Yet soon the idea of space travel was dealt two almost fatal blows: the maps of Hevelius illustrated a dead moon in 1647, and Cassini discovered that the distance between the sun and earth exceeded 80 million miles. "Things had suddenly grown too large, too impressive, for light-headed speculation on actual travel," Ley concluded.[25]

Fortunately the nineteenth century included "the decades of great dreams." Ley connected the history of astronomy to the history of newsprint, which continuously excited readers with hoaxes and astronomical wonders. The most dramatic was Sir John Herschel's alleged discovery of vegetation, unicorns, and "bat-men and bat-women" on the moon. The hoax had been printed in 1835 in the New York daily, the *Sun*, which saw its circulation rise to 19,300 copies, far more than the *London Times*. Astronomical discoveries could be profitable for newspapers. "For a while New Yorkers were indignant," Ley wrote, "then they began to laugh . . . and the 'Panorama' exhibits and the stage began capitalizing on . . . [the] great moon hoax." Then came the actual astronomical discoveries during the late nineteenth century: a primordial Venus and an advanced Mars, crisscrossed by strange canals. Suddenly the nineteenth century became the era of grand astronomical dreams. Ley clearly enjoyed dissecting the competing hypotheses of life on other worlds, while discussing the evidence. At times he distinguished between crackpots and astronomers. Nevertheless, popular science and popular literature complemented each other. Ley commented on "three decades of Mars enthusiasm, three decades during which reports from astronomical observation were awaited and read as avidly as reports from the front in the middle of a war." He added,

"It goes without saying that literature did not fail to contribute to these decades of the great dreams."[26]

In the early twentieth century new astronomical discoveries, as well as the debunking of earlier theories, discouraged many of these dreams. By the 1920s Mars was "out of the running" and Venus was "somewhat too warm" when it came to intelligent life. Still, the dream of the human conquest of space lived on, with imagined journeys to nearby worlds. Newer novelists were "confronted" with scientific evidence, forcing them to "let the conquest of space originate from earth." "The chances are overwhelming," Ley argued, "that future developments will prove them correct."[27]

The remainder of the book presented a detailed history and discussion of rockets as well as a biographical account of the VfR. In clear and direct language, Ley explained the how, why, and when of rockets, from ancient Chinese powder rockets to liquid-fuel designs of the late 1930s. He also evaluated the contributions of different theorists and engineers, with the aim of establishing a clear chain of events regarding the development and progression of the field. The most dramatic chapters are called "The Battle of the Formulae" and "Success, Failure, and Politics," which read like a tell-all history of the VfR, Rocket Port Berlin, and the decline of experimentation in Germany. Overall, the rise of the Nazis killed the scientific movement, forcing Ley and others to flee Germany. In a footnote he even claims, "I have it in writing from his own hand that . . . [Oberth] denounced me to his Nazi superior, stressing the fact that I was in correspondence with Ziolkovsky, Rynin, and Dr. Perelman." For future innovation in rocketry, politics had to get out of the way of serious engineering.[28]

The final chapters of the book discuss possible future innovations of rockets, from meteorological instruments to cosmic voyages. The discussion concludes with an answer to the question of "Why should we try for space travel?" The "simple answer" is presented: "Somebody has got to start at some time, and we may as well get the glory for our own century." Ley then reassured readers: "It can be added that developments of this type very often progress much more smoothly than expected as soon as the initial difficulties have been overcome. . . . We know about them and we know *when* they will be overcome." Ley also stressed that the costs of the first journey would be offset by the sale of lunar material. Anything lunar, he predicted, "will bring fabulous prices . . . but only once." While the ultimate payout is the knowledge of discovery, Ley further predicted that the discoveries made on a station in space could possibly "pay for everything."[29]

Initially Viking Press did not promote the book heavily, beyond a few small advertisements in newspapers. The book met with an enthusiastic, albeit limited reception. Most notably, science writer Waldemar Kaempffert reviewed it for the *New York Times*. "Though the head of Willy Ley may be somewhere in interstellar space," Kaempffert asserted, "his feet are on the earth." While Kaempffert commended the thoroughness of the book, he took issue with Ley's optimism. Kaempffert also critiqued Ley's interpretation of the history of science, implying that it was absurd to attribute the lack of medieval scientific progress to an absence of skepticism in the world. Nevertheless, Kaempffert and others kindly reviewed the merits of the text, while sharing in Ley's excitement for the future. *Newsweek* even remarked on Ley's obsession with rockets: "Scoffing didn't bother Ley. His thinking was miles ahead of actual rocket development then or now, but his predictions were sublimely confident." Otherwise the book had a fairly limited reception. Astronomer Robert S. Richardson reviewed it in *Publications of the Astronomical Society of the Pacific*, calling the rocket "perhaps the most exciting machine in modern science." Overall this reviewer described the book as "a masterful and fascinating account." The technology inspired awe and wonder.[30]

Robots and Rumors

Ley continued to write articles on traditional war weapons for *PM* and *Mechanix Illustrated*. Given the broader context, newsprint coverage focused less on technologies of the future and more on the mysteries surrounding new "robot bombs" that fell mostly on London, from June 1944. Ley's *Rockets* may indirectly have influenced reporters to speculate about Hermann Oberth's role as "the man who perfected Hitler's flying bomb." Other reports included striking front-page headlines, such as "Robots Kill 2,752," "Germans Unleash Gigantic New Robots," and "Robots Rain on Britain All Day in Greatest Sustained Attack Yet." In reality, the V-1 "buzz bomb" was an ineffective weapon.[31]

Although Ley was as curious as other reporters, he tried to calm the readers of Mechanix Illustrated with an article titled "The Future of the Robot Bomb." He probably wrote this piece in early September 1944, at exactly the moment when V-2 rockets began to hit London and Paris. Ley seems quite unaware of reports about these new rockets. Likely, any reports were conflicting, often filed under the heading of never-ending rumors of Hitler's secret weapon. Ley based much of his evaluation on existing accounts of V-1s. Ley began, "The German flying bomb is a weapon of paradoxes." In a sharp tone, he wrote: "It is novel, yet the idea is by no means new. It is crude and inaccurate, yet it is destructive. It has no military value in the present war and will not delay the fall of Berlin by a single day, yet it may be the dominant weapon of future wars." Although the robot bombs were simply "too new" and "not quite out of the laboratory stage," they could serve as the "means of preventing any future German aggression." Ley also speculated on the ways in which the V-1 could be improved through enlargement and the elimination of vulnerable launch sites. With an increased frequency of launches, these "very heavy and . . . very long range" aerial torpedoes might become dominant technologies of war. "Naturally they would not win the war by themselves, not any more than any other weapon," Ley argued, "but they might be dominant in about the same sense as the airplane is the dominant weapon of the present war: it cannot win by itself, but you cannot do without it." These thoughts led Ley to the final paradox:

> While full of future possibilities the robot bombs of the Nazis did not accomplish anything from the military point of view . . . this weapon may see to it that future generations of Nazis cannot start another war. . . . The flying bomb in itself might be enough of a threat . . . and those Germans, who, in the future, do plan war, will have to console themselves with the thought that they cannot make war because of a German invention.[32]

One thing was certain: by September 1944 the German war effort was a hopeless campaign against inevitable defeat. No wonder weapon would save the Nazis. Ley must have felt a sense of pride at his role as a scientific educator and debunker of propaganda. He must have also felt a sense of closure, as he left *PM* and started a new phase of his life as an aspiring rocket engineer, first in Atlanta, Georgia, and then in Washington, D.C. He moved his family at exactly the moment when V-2 rockets began hitting London. This phase is the focus of the next chapter. At the time of the move, Ley's life had been defined by a broader crusade. He was not alone in his endeavors.

The Science Writers and the Second World War

Ley belonged to a broader New York community of freelancers, scientific historians, and editors who had enlisted in a fight against totalitarianism, public hysteria, and irrationalism. In order to place Ley's writings and activities into this broader context, it is necessary to examine the commonalities between his efforts to educate and a popular crusade that united many different scientific intellectuals. Taking a retrospective look at the careers and activities of other writers and historians of science, allows the legacy of Ley's *PM* years to become obvious. It requires us to take a step back and view the overall scene.

When war engulfed Europe, some historians of science restated their cause. In particular, founding father George Sarton announced in the journal *Isis*: "In the face of the moral and social chaos endangering the whole world it is more necessary than ever to study . . . our most precious heritage, the heritage not of one nation but of the whole mankind." *Isis*, as a journal devoted to the history of science, was not simply aimed at a better understanding of past contributions of scientists. He explained: "The purpose of *Isis* is to explain our past efforts in that direction and thus help to continue them in the same spirit of devotion to truth and humanity." Scientists, as witnessed in the past, could be crusaders for this cause. They could proclaim truth and save humanity. The historian of science could further celebrate the past crusaders who cast light upon the darkness. "In the shadow of so many crimes and sufferings," Sarton elaborated, "can there be a greater consolation than to study and explain more clearly the best and highest deeds of the people of every nation and thus to vindicate the goodness of man?"[33]

Other historians and scientists advocated a more direct route. Anthropologist M. F. Ashley Montagu wrote: "It is untrue that science is unpolitical. Everything is political." Montagu exemplified a scientist who embraced a crusade to educate a mass audience through popular writing. Scientists must "convince their fellows of the value of the contribution which they have to offer." They must enlighten the society in which they live. Only under their guidance and direction can that society progress toward a better, safer, and more rational future. Montagu's campaign to educate, enlighten, and liberate American minds can be read in his most popular book, *Man's Most Dangerous Myth: The Fallacy of Race* (1942). The book was a no-holds-barred attack on Nazi pseudoscience and American racism. Science, Montagu asserted, proved "the fundamental unity of all mankind." The diseases of bigotry, intolerance, and racism must be purged from the body politic. Scientific thinking served as a cure for a tyranny of tradition that "must be broken." As a reviewer observed, Montagu's writings attempted "not only to popularize but also to re-educate morally."[34]

Other scientific writers echoed these sentiments. As described by historian David A. Hollinger, this camp of diverse New York intellectuals participated in a broader cause. From philosopher Karl Popper to cultural anthropologist Margaret Mead, these intellectuals participated in a cultural struggle, defending science and democracy in a shared endeavor to save the public from both foreign and domestic threats. Many of these intellectuals waged a war that defended democracy and science against the spread of a generalized totalitarianism. They contrasted the open, public, and democratic aspects of science to the closed worlds of authoritarianism. A rational democracy relied on the type of thinking that promoted open-mindedness, critical thought, objective evaluation, and informed consent. In their perspective science relied on democratic values and practices. Conversely, totalitarianism relied on obedience and a closed system.

Science writer Waldemar Kaempffert made similar claims. Just as Ley educated the readers of *PM*, Kaempffert worked diligently for the *New York Times*, in which he sought to educate and enlighten the general public. One could even read Kaempffert's 1956 obituary as a shared description of Ley: "He wrote thousands of articles to inform the average reader . . . without bewildering the reader by talking over his head or patronizing him by talking down. Though he often dramatized science, he never tried to sensationalize it." Kaempffert's articles are quite revealing. For example, his January 1941 essay titled "Science in the Totalitarian State" began by lumping communism and Nazism together. "Both agree," Kaempffert wrote, "that the university professor must serve the state, accept the tenets of official ideology and eschew any excursions into the metaphysical or the theoretical." He continued, "Objectivity is derided in both the Soviet Union and Germany as unattainable and as anti-social." Both regimes sought to crush individuality and impose official dogma. Both systems sought to eliminate freedom of inquiry and dissent. Under such constraints, academic freedom was impossible.[35]

According to Kaempffert, ideological blinders led Soviet and German scientists to denounce certain theories. Anti-Semites denounced the theory of relativity as "an example of characteristically perverse Jewish thinking." In Soviet Russia relativity was similarly denounced as "an expression of 'bourgeois idealism.'" Kaempffert then lists several other examples of the ideological manipulation of science in accord with the irrational dogma of the state and its occult teachings. In these settings, the first criterion of science is not the truth. Rather, the first criterion is the theory's compatibility with ideology. Kaempffert listed the ways in which ideology, vehemence, and sheer blood lust had deeply perverted Soviet and German science. He then pleaded for the preservation of science and democracy by equating the two activities: "[If] the dictators are to be overthrown, if democracy is to be preserved, the part that science and technology played in the rise of democracy cannot be ignored."[36]

The conclusion is clear: there can be no compromise between science and the totalitarian mindset. No self-respecting scientist can function in a closed society. The totalitarians may have taken control of machinery, organizations, and research centers. However, in Kaempffert's view, there is nothing that they can do to preserve the scientific attitude of mind in their regimes. Thus, "there can be no Newtons, no Darwins, no Einsteins." Science could only progress in "the fundamental freedom of democracy." "There can be science and engineering under dictation," he summarized, "but it will be stylized science, engineering which does not progress."[37]

For the remainder of the war Kaempffert's "Science in the News" articles made similar ideological points, while other articles took his case further. Kaempffert relentlessly campaigned for the mobilization of the scientific community. He complimented the scientists, public educators, and national leaders who served on the frontlines of a war to preserve human freedom, democracy, and science. Like Ley, he informed citizens, while simultaneously training them to think scientifically about war and its dangers.[38]

Other science writers contributed to the scene. Notably, Howard W. Blakeslee of the Associated Press served as an influential expert, often educating Americans about the technologies of war and the science behind future applications. Blakeslee forecast amazing postwar advances that followed the mobilization of the scientific community. Other AP science writers, such as Frank Carey and Rennie Taylor, shared his enthusiasm for the postwar future. John J. O'Neill led a similar crusade for scientific education in the pages of the *New York Herald Tribune*. As both science editor and president of the National Association of Science Writers, O'Neill campaigned for the mobilization of the scientific community. Yet O'Neill warned that the Roosevelt administration was "staging a totalitarian revolution against the American people." After stating that scientists had recently discovered a method of releasing large amounts of energy from a single atom, he argued, "Can we trust our politicians and war makers with a weapon like that? The answer is no. Nevertheless, our politicians have taken control of the scientists." In O'Neill's perspective, the imposition of secrecy and state-directed science was inimical to the relationship between science and democracy.[39]

While the science writers educated Americans and debunked myths and propaganda, other scientific intellectuals came together to voice their perspectives. For example, a significant 1943 conference in New York City included various professors, scientific leaders, and directors of institutions. Their papers were published in *The Scientific Spirit and Democratic Faith* (1944). According to Editor Eduard C. Lindeman, the conference and the publication served as "a clear protest" against totalitarianism. Understanding, exploring, and celebrating the "essential interrelation" of democracy and science became a key to fighting the spread of both authoritarianism and irrationalism. Lindeman explained:

> Indeed, a new authoritarian movement, almost a coalition although not consciously organized, had arisen in our midst. Strange voices using masked words were heard throughout the land, voices demanding a new authority in education, in morals, and in government. These voices used the familiar words of the democratic tradition but the ideas were not of that tradition.[40]

Overall the authoritarians "asked for allegiance to fixed principles, inflexible rules of morality, and unquestioned acceptance of a supernatural interpretation of human experience." Lindeman then blamed intellectual elites "in positions of power and influence," religious fundamentalists who "found themselves again in ascendancy," and mass-media publications that profited from the spread of such nonsense. Lindeman asked: "How could such divergent groups find a common denominator, the sophisticates and the illiterates, rich and poor, the powerful and the defeated?" The answer is stated as follows: "They held in common their fear of the future, their anti-scientific preconceptions . . . their terrible need for certainty and security, and an identical logic."

Unlike the disparate groups that supported fascism, the contributors to this conference believed in the dynamic power of science and truth. They celebrated a synthesis, which "combines science as a search for truth, democracy as the guarantee of liberty, humanism as the source of faith, and education as the instrument of progress." The last part of this statement could not be overstated. Education was crucial in the battle for hearts and minds. Public outreach was not an attempt to counter one dogma with another. The scientific spirit involved critical reasoning, the careful evaluation of evidence, and a solid tradition of anti-authoritarian bravery. The interrelations of science and democracy became the most important American tradition in need of preservation.[41]

It was not enough for these journalists, scientists, and intellectuals to assume prominent positions of power. They also had to communicate the scientific spirit through education and public enlightenment. Obviously many of these sentiments stemmed from a much older discourse among scientists and academics. In fact, as historian Andrew Jewett has noted, much of the equivocation of science and democracy can be found in nineteenth-century discourse. Long before the rise of technoscience or big science, there was a massive effort for the mobilization of scientists and the preservation of American democracy. Just as writers of popular science owed a great deal to nineteenth-century authors such as Richard and Mary Proctor, science journalists owed a great deal to a broader intellectual tradition.[42]

In the perspective of many thinkers, not only did science offer a rational and fulfilling path—scientific thinking also served as an antidote against the most unfulfilling, dangerous, and cultist aspects of state ideologies. Totalitarianism embodied the past and its horrors. In this worldview, the fascists and the communists were medieval in mentality. Democracy embodied the future, its wonders, and most important, its freedom of thought. Such a crusade for the present and future required disciples who were unafraid of a newspaper reporter or a microphone. It needed orators and writers to rush to the bully pulpit and preach the scientific gospel. It needed great communicators who could save souls from falsehoods, irrational beliefs, and cultist nonsense. Jewett compared these activities to a "missionary enterprise." Such language is appropriate.

Postwar Dreams

Willy Ley had spent most of the war earning a precarious living as an evangelist for the scientific spirit. He had worked hard to educate Americans, combat hysterical notions, and serve the war cause in his own way. He must have felt a degree of pride in his accomplishments as a science writer and technology expert. Whereas many newspapers and tabloids had profited from public fears and anxieties, Ley's *PM* articles sought to calm readers and bolster their faith in Allied technologies of war. Ley had debunked the nonsense and dethroned several of the phonies. In many ways, he had taught his readers to think scientifically about the dangers and realities of war.

By late 1944 the war was coming to an end. Ley had no intention of continuing his career mainly as a freelance writer, living on book royalties and small checks. As a new father, he needed a more reliable and stable income. While the public clamored for more information on Hitler's mysterious weapon, he left New York for Atlanta, Georgia. Finally, Ley would become an American rocket engineer, or so he hoped.

An Engineer's Postwar Dreams

Throughout September and October 1944 news reports and rumors circulated. A mysterious weapon rained down upon London, and eye-witnesses came forward. As journalists scrambled for information, Ley moved his family from New York City to Atlanta, Georgia. He joined the newly formed Burke Aircraft Corporation. Very little is known about his work or the general plans of the petitioners for incorpora-tion: Arthur J. Burke, H. Eliss, and H. Flynn. The company planned to "design, manufacture, and sell power plants for the operation of heavier-than-air craft" as well as "meteorological or 'coast guard' type life saving [sic] rockets, signal rockets and jet propelled, rocket type projectiles."[1]

Although Ley's motivations for the move to Atlanta are undocu-mented, it seems likely that he perceived an opportune moment to position himself as America's rocket expert, capable of applying his expertise to an emerging field. Ley realized that the time was ripe for American rocketry. The United States needed a rocket expert. In his mind, he was the most qualified individual for the task. For the next four years, Ley battled for governmental contracts, first in Atlanta and then in Washington, D.C., when Burke Aircraft Company became ab-sorbed in the Washington Institute of Technology. While he tried to establish himself as a competent engineer, he wrote fewer articles for magazines. His research and writing decreased.

Ultimately Ley did not succeed in securing a contract. The U.S. Army and Navy excluded Ley from the centers of rocket research. Instead, they increasingly relied on captured Germans who had worked in Nazi research centers. Ley's attempt to transition from a science writer to an engineer ended in bitter disappointment. By 1948 he had recommitted himself to writing popular books. He had gambled and lost. For the rest of his life, he would be an outsider to the centers of research and development. Yet he also became America's rocket expert, before engineer Wernher von Braun took center stage.

The Shock of V-2

Back in September 1944 Ley had received a visit from A. V. Cleaver, an aviation businessman and member of the British Interplanetary Society. Cleaver had classified knowledge about new weapons that hit London: large rocket missiles. Cleaver spoke about public rumors, without divulging intelligence. Ley dismissed the rumors entirely. Cleaver recalled: "I was astonished to find that, for some reason, he had decided that the rumours were a lot of nonsense. . . . I could describe the rocket to him if he would only listen!" Ley did not believe reports of V-2 rockets until November, when both Churchill and Hitler spoke publicly. Even then Ley had many doubts about conflicting reports, rumors, and eyewitness accounts. How could Ley, as a prophet of the Space Age, have been so blindsided by the V-2? The question is not difficult to answer, given his expertise on war weapons. He had many reasons to doubt the accuracy and effectiveness of a long-range missile in 1944. It must have seemed as though the Nazis turned to fire arrows amid fire bombings. In his judgment, no known payload could justify the military value of war rockets.[2]

Most likely Ley received requests for articles, yet he did not immediately comment. A notable exception occurred in the January 1945 edition of *Technology Review*. Ley argued, "The military value of this weapon is small . . . any hit scored by the V-2 is purely accidental and completely unpredictable." Ley added, "The effect of the one-ton war head is actually less than that of the same war head when attached to the V-1." He concluded: "V-2, therefore can be characterized as an extraordinary example of engineering and research but also as a military flop . . . V-2 lacks accuracy, completely." Ley may have expressed similar views in interviews with a local radio station.[3]

Aside from this short article, Ley became reluctant to comment on the V-2. He spent many weeks simply collecting information. Then he produced an exclusive piece for *Astounding Science Fiction*. The editor introduced the article, "V-2 Rocket Cargo Ship," by claiming, "Willy Ley knows rockets—and German rocket engineers. He can, and does, identify the man who designed V-2." Ley first asserted, "The full story of the German rocket research laboratory near Peenemünde . . . will never be written. There will be nobody alive who can write it." Ley believed that the researchers would be shot by the Nazis: "We cannot hope to take Peenemünde. . . . The Nazis will see to it that everything will be utterly destroyed before we get there. And Himmler, I am sure, has lists. . . . If they escape future Allied bombings, they will be shot by the Gestapo." In retrospect the article demonstrates Ley's lack of insider knowledge, not only about Peenemünde but also about von Braun's earlier work and the Army Ordnance's support for experimentation. The piece also discredits Cleaver's story, in which Ley believed that Wernher von Braun was the man behind the V-2 rocket. For example, a different letter from Cleaver to von Braun stated: "[Ley] then said that, if the rumors were true, 'a young man called von Braun' might be responsible."[4]

In the *Astounding* article, Ley recalled, "At the time Hitler was actually coming to power no rocket research went on anywhere in Germany and this state of affairs was to prevail for another three years." This statement conflicted with his 1944 account, in which the army militarized rocket research. Ley now believed that research stagnated until 1937, when Oberth "established . . . contact with the Germany Army." Soon, at Peenemünde, "Oberth . . . probably was the department head of the V-2 branch." Ley then recounted newsprint stories and other information about the V-2, its design, and its fuel. "Everything about it spells out OBERTH in capital letters," Ley concluded.

Lastly, he argued, "Barring miracles we will not be able to continue for peaceful purposes what the Germans started with war in mind. But . . . Peenemünde proved that it can be done." Ley does not mention von Braun, the person most responsible for the V-2 rocket.[5]

Ley positioned himself as the man who could recreate the successes of the V-2 rocket. For example, in late spring 1945 he spoke publicly about his meteorological rockets. Apparently he had made much progress with a motor designed for a ten-foot rocket. He claimed that it would soar 2,000 feet per second up to 85,000 feet, before parachuting down with instruments and data unharmed. Ley even speculated about the cost benefits of future mass production of meteorological rockets. He was optimistic about the future. Ley also continued to give interviews with local radio stations. Yet the move to Atlanta had been a mistake. Burke Aircraft proved to be a dead end. By June 1945 Ley grew disgusted with both the climate and his employer. He tentatively resigned from Burke by airmail. He explained to Heinlein, "I'm in no mood for further nonsense, four months breach of contract is enough." Ley gave Burke a deadline of June 30. "If by that time," he explained, "finances are straightened out to satisfy me, I promised to make a new agreement." Ley planned to move back to New York City, where he would accept an offer to become an editor for *Mechanix Illustrated*. He looked forward to the move: "If things go bad . . . I have to go on here [in Atlanta]."[6]

This statement inaugurated a long period of frustration with organizations, contracts, and tentative connections to branches of the military. For the next three years Ley would act as an aspiring engineer and expert who was in the dark and on the outside, desperately trying to convince officials that his knowledge could be utilized. He gradually came to terms with enormous obstacles that prevented him from gaining insider status. He also discovered that many of the Peenemünders had not been killed. In fact, the former Nazi engineers were now working for the U.S. military. Not only would Ley come to realize his disconnect from the centers of research, but he also realized that his own livelihood as an American expert on rocketry depended upon a cordial relationship with the former servants of the Third Reich. Eventually he embraced his role as an outsider, and he reconciled his conflicting views of the ex-Peenemünders. He also came to terms with the military and political realities of the Cold War, which gave new life to the rocket as a weapon of war.

Early Hopes of an Aspiring Expert

In the final days of the war Ley searched for an escape from Burke Aircraft. He reached out to his friends who had connections with the military. Most directly, he courted Heinlein and his relations with navy superiors. A series of rather cryptic letters between Ley and Heinlein present a partial account of what followed. In June 1945 Ley traveled to New York, where he attended a meeting. Ley described the event to Heinlein:

> I had a conference lasting about two hours, partly with your friend, partly with a specialist from the proper projectile section who was called over. In these two hours we talked the problem over from all angles, and it seems to me that I gave satisfactory answers and outlined satisfactory plans. Of necessity everybody was a bit tight-lipped, but we got along fine and unless I get a report to the contrary I'm going to think that the meeting was successful.[7]

It is unclear if this meeting was related to the events that followed. On the Fourth of July Willy and Olga drank champagne. They did not celebrate a return to New York City. Instead, the Leys planned to move to Washington, D.C., where Ley would become an employee of the Washington Institute of Technology, after it absorbed Burke Aircraft. Although WIT specialized in radio technology and plastics, it aimed to get into the field of aeronautics. Ley would lead its efforts in meteorological rockets. He explained: "I am paid $500 a month for the purpose of carrying out the necessary groundtests [sic] for rocket motors which are to go into meteorological rockets." Whereas the Ley of earlier years saw meteorological rockets as the next logical step, he now argued, "I feel that meteorological rockets are slightly ridiculous after V-2 . . . [but] they may be something of commercial value."[8]

Ley explained how the WIT merger came about in a previous letter to Heinlein. After his New York trip, a local banker asked him to go to

Washington. This man agreed to pay all of Ley's expenses. The banker indicated that his "friends . . . might take Burke Aircraft over, pending a discussion." It may have helped that Ley owned many shares of Burke, given to compensate him for unpaid work. Ley recalled: "I went with great misgiving, calling myself an idiot for going to the trouble of a trip to Washington for nothing . . . well the result was a kind of temporary merger." The WIT agreed to assume all responsibilities for Burke Aircraft until January 31, 1946. This was Ley's first deadline to secure a government contract for meteorological rockets. If he failed in that task, Burke and WIT would part ways. Ley was optimistic: "Naturally, this is a great step forwards, the WIT has extensive laboratories and facilities, they employ a complete staff of specialists in all kinds of fields . . . so the work should make rapid progress." He also looked forward to the job, adding, "My presence in Washington might also be useful in other respects."[9]

Heinlein continued to work behind the scenes to facilitate Ley's contact with Captain Cal Laning of the navy. Ley first spoke of his relationship with Laning in retrospective and disappointing terms: "My conversation with Captain Laning led me to expect that after the war I would be put in a position, financially I mean, to carry on the necessary experiment for a Moon Messenger." Ley continued: "I was given to understand that certain things would have to be kept secret since the Navy was interested in long distance rockets of the V-2 type and that it was naturally understood that any of my work which could be applied would be applied." Ley also summarized a previous meeting with one of Laning's subordinates, telling Heinlein, "I am at a complete loss to judge what is going on." Ley added: "I know that I can keep a secret if I am told that it is one, the question is therefore very simply of whether they are going to trust me or not."[10]

Ley moved to Washington in late September 1945, after the surprise ending of the Pacific war. By October Heinlein offered apologetic excuses for Laning's earlier treatment of Ley, arguing that Laning was upset and frustrated by the necessary secrecy. In November Ley updated Heinlein on his relations with Laning: "Meanwhile Captain Laning came around twice. . . . We are getting along beautifully. . . . L. wants me to get together with some of his men, just as soon as restrictions

have been lifted enough to enable them to talk freely." Ley then updated Heinlein on his progress at WIT: "By the time you get this letter
my teststand [sic] will be fully assembled and tested and IF no leaking
welding seams are found—I DON'T TRUST WELDING SEAMS!!!"[11]

Ley was still optimistic, yet he was in the dark and on the outside.
Like most Americans, he had no knowledge of Operation Paperclip
and other efforts to recruit and transport German scientists to the
United States. By the time Ley responded to Heinlein, the first group
of seven German engineers had already arrived at Fort Strong, in Boston Harbor. Wernher von Braun was among them. Eventually other
colleagues would arrive at Fort Bliss, Texas, and the White Sands Proving Ground in New Mexico.

Atomic Realities

In an earlier letter Heinlein launched into a heartfelt plea for the internationalization of atomics. One might get the impression that Heinlein prodded Ley, after Ley's factual article "Inside the Atom" appeared
in Natural History. H. L. Shapiro of the American Museum of Natural
History introduced the article. Shapiro stated, "We cannot hope to remain the sole guardian of the scientific knowledge that has made the
atomic bomb possible." Consequently, "it is the solemn obligation of
scientists, particularly those concerned with human affairs, and of all
men of vision, to work for a political and social organization of mankind that will, by making war impossible, permit us to employ our
powers without disaster."[12]

Ley responded to the "awful question of atomic warfare." In a long
and revealing passage, Ley agreed with Heinlein's internationalism
after stating, "I am as worried about it as you are . . . and I cannot
find a way out of the moral labyrinth." "So far," Ley continued, "Congress AND the senator from Missouri who is president (if only the
real president were still alive) have done everything possible to make
an atomic war certain." Ley added his thoughts on the role of the scientist in the age of nuclear weapons: "But the era has come where the
world listens to the scientists. . . . The picture is not all black, there
is still hope." Ley concluded, "Yes, I agree in principle with complete

internationalization of atomics and I also agree in principle with relinquishing sovereignty. . . . I'd rather give up sovereignty than disperse, I am not too certain that I would enjoy life if there are no cities left." The last sentence of the letter contained a permanent goodbye to Atlanta: "We still would want to test dangerous atomic reactions. . . . There is a perfect testing ground not far away: the six south-eastern states!"[13]

Ley voiced similar sentiments in *Mechanix Illustrated* about atomic war. His articles contained a mixture of hope, fear, and internationalism. For example, in an article called "Peace or Else!" he wrote: "Earth has become a world of Either/Or. . . . Either—we are firmly determined that there shall be no war, and spend as much energy, thought and money on the problem of preventing it as we now spend in preparing for it. . . . Or—we blunder into irretrievable errors." If the latter path were chosen, atomic warfare was inevitable. It would "come without warning and without declaration, with a fury so monstrous that the mind cannot conceive it." However, if humanity could follow a path of peace, the future might include a long list of technological and scientific achievements as well as a period "of unlimited progress, of infinite riches of knowledge and material riches, of immediate preliminaries to humanity's spread through the solar system as a first step to a spread through the galaxy." Ley saw the situation in dire terms. Scientifically, the destructive potential of atomic weapons was "so incredible that the mind rejects it." "The deaths of 100,000 Japanese in five microseconds," he argued, was "*the gentlest possible application of the smallest possible atomic bomb.*" Unlike in previous wars, in which ground defenses could counter aerial bombardment, there were no effective counter-measures to atomic bombs. Suddenly, the war rocket had an utterly devastating use: "Couple V-2 and the atomic bomb (it can be done today) and you have a destroyer of cities against which there is no defense once the rocket is in the stratosphere." He concluded, "We need to realize that a new era in human relations is here, an era that no longer permits the concept of war which now means *complete, mutual,* atomic destruction. What we have to learn is to live with atomic energy!"[14]

More Rockets and Rumors

On December 17, 1945, Ley attended the Wright Brothers Lecture at the Chamber of Commerce in Washington, D.C. An expert shared detailed information about British jet engines. Ley recalled: "I began to feel a vague yearning for simplicity. Several compression stages and several turbine stages, with six or eight combustion chambers twisted around each other began to look mildly frightening after a while." He left the talk slightly "confused and somewhat bewildered" to stroll down a Washington avenue. Slowly, Ley approached an outdoor navy exhibit that included a V-2 rocket, which he had only seen in photographs and newsreels. He remembered: "Meanwhile it had grown quite dark and a perfectly round moon rose in the East. It was purely an accident, but the big 46-foot rocket, lying sloping on the bridge section truck, raising its nose some seven or eight degrees, pointed directly to the rising moon. It looked 'target for tonight.'" A sudden emotion overtook Ley. He remembered: "One should be immune against sudden thoughts which spring up on such occasions; but . . . almost without any volition on my own part I said: 'Still in our lifetime.'" He saw the V-2 as an embryonic spaceship.[15]

By spring 1946 Ley discovered that many of the former German rocket engineers had been captured by the United States Army. He may first have suspected this after a War Department press release in October 1945 confirmed the transport of "certain outstanding German scientists and technicians." It is unknown how Ley learned that Peenemünders were among this group. This information contributed to his increasing frustration. On May 5, 1946, he received a letter from Laning, which read, "Dear Willy, we aren't having much luck getting you into the Navy rocket picture." Two days later, Ley voiced his anger in a letter to Heinlein. "So far, unfortunately," he wrote, "nothing has worked out. I was supposed to translate German documents,— no soap. I was supposed to interrogate the captured German rocket experts—impossible. I am not invited anywhere for anything." He added: "Apparently there is some higher-up bozo somewhere who does not want to deal with German-born citizens, but prefers to deal with genuine captured Nazis instead, presumably because they are, at least, not civilians." Ley went on to predict "a big disappointment . . . about White Sands."[16]

The situation was not as bleak as Ley perceived. A week or so later, a meeting at the Navy Department altered Ley's situation. He wrote quickly to Heinlein to recant his complaint, adding, "It seems that a big contract is coming up, big at least for a firm the size of W.I.T." Under this arrangement, WIT would present a proposal and another conference would take place. If the navy agreed, then WIT's business manager would finalize the details. A few weeks later Ley still expressed his confidence, although he grew baffled by navy bureaucracy. Yet all signs pointed to a contract:

> Well, now I am sitting here waiting for the contract, there is some re-writing going on in the Navy Department and that re-writing (hold your breath) contains the provision that, a few months hence, the project is to be declassified . . . and that W.I.T. is THEN EXPECTED TO SEEK PUBLICITY for this project. Friends, brothers, colleagues and toverishtshi [comrades] . . . I rejoice, my heart is happy, but I wish I could understand the Navy![17]

By August 1946 Ley's disgust with navy bureaucracy returned. He told Heinlein, "Then, of course, there is our friend, the Navy. I am right now making estimates, crystal gazing at its worst and sometimes I feel like looking a reference up in Nostradamus." He continued: "The official request for a bid is in . . . and what they crammed into the specifications would have kept Peenemünde busy for a year." He also complained about the rising influence of Wernher von Braun, who did not endorse Ley for a contract. Ley recalled: "When L. asked von Braun whether he knew me von B simply said 'yes' and fell silent. I wanted to know what would be done with him, but L. did not know." Ley added, "I only hope that the U.S. Army will not suddenly find him 'charming' in addition to being useful." He then complained about a recent announcement by the army to launch a guided missile to the moon within eighteen months. Ley asked, "Just who is writing science fiction these days?"[18]

In September 1946 Ley still felt optimistic. "The bid is in," he wrote, and "my own reasoning seems to have official support, namely . . . don't

take something that was developed for another purpose (long range artillery) just because it is bigger and try to go from there." The V-2 did
not provide a blueprint for the future. It was a rocket of war. Ley stood
poised to offer a rocket of peaceful, scientific exploration, if he could
procure the necessary funding.[19]

The Struggle against Nonsense

While Ley continued to express doubts as to the potential of the V-2,
articles began to appear in the press that argued otherwise. Most notable was G. Edward Pendray's "Next Step the Moon," which appeared
in Collier's. The article focused, in part, on the army's moon announcement, while claiming: "Such a rocket would be only two or three times
the size of the 14-ton German V-2 rockets of World War II." For Pendray, a moon shot with sophisticated instruments might be possible by
1952. This article was Pendray's third contribution to Collier's, which
had already published space-related articles.[20]

Pendray had gained momentum in the press as a rocket expert. This
momentum had begun as early as 1930. By 1946 Pendray cultivated the
persona of a "Yankee Rocketeer" as a wild-eyed visionary. He published
articles in Coronet, Harper's, Collier's, and other popular magazines. For
example, in 1945 Popular Science carried his "The Reaction Engine."
It must have astonished Ley that the magazine introduced the article
with these words: "This authoritative article is the first attempt in any
language to show the relation of the various kinds of rocket power now
in use." Pendray's claims most likely irked Ley. For example, in 1943
Pendray hinted at secret governmental programs that made the rocket
"one of the most successful weapons of modern warfare." Ley would
have described that statement as complete and utter nonsense.[21]

Much of Pendray's success followed the publication of his book,
The Coming Age of Rocket Power (1945). The work lacked Ley's style of
personal memoir as well as his historical framework. Nevertheless,
it succeeded in translating complex concepts for the general reader.
Ley praised the book, saying, "Mr. G. Edward Pendray . . . has written
a book on his favorite topic which will not only be interesting reading for the layman but may also serve well as an introduction to this

field for engineers." Nevertheless, Ley criticized Pendray for his many errors: "But while he proves himself a master of the broad stroke in painting this complicated picture, Mr. Pendray's attention to detail is not meticulous enough to make his book a reference work." Ley also claimed, "The book falls off in a sharp curve when it comes to predictions." On some matters Pendray was consistently pessimistic, yet on other matters he advanced a "curiously strained and forcefully optimistic discussion." Despite these flaws, Ley congratulated Pendray on his "first primer." Ley said nothing about *Harper's* promotion of the book, which labeled Pendray as "an outstanding world authority on rockets and jet propulsion." However, a reviewer casually remarked, "As far as we are concerned . . . we can think of numerous other persons who could probably share our sentiments." Other reviewers were less kind to Pendray. Lionel S. Marks of the *Scientific Monthly* claimed, "The book is primarily a statement of what a rocketeer feels that the rocket will be able to do rather than an attempt to ascertain its foreseeable developments in the not-too-distant future."[22]

Normally, Ley did not view his colleagues as competitors. He praised later books like Arthur C. Clarke's *The Exploration of Space* (1951). Ley applauded most works that furthered the cause. Yet with Pendray, the situation was different. Ley viewed Pendray's focus on passenger rockets and earthbound travel as detrimental to the cause of space travel. It is easy to imagine Ley scoffing at Pendray's statement: "For myself I do not know whether rocket power will ever permit fulfillment of our ambitious desire to reach the moon. Perhaps it isn't of very much moment, for in the age of rocket power jet propulsion will find plenty to do right here." Ley commented on Pendray's pessimism, yet he did not comment on the book's agenda, which highlighted the contributions of Robert H. Goddard and the work of the American Rocket Society, while characterizing German developments as, according to one reviewer, "the Fritz Opel publicity-seeking variety." Ley may have viewed the book as a personal insult to Oberth, the VfR, and the engineers who designed the V-2.[23]

The text also contained a few outlandish assertions. Pendray implied that the German rocketry fad began with publicity surrounding Goddard, particularly after Oberth "received a copy of Goddard's 1919 report directly from the author." Pendray added, "By 1923—the year Oberth's book appeared in Europe—Goddard had reached the point of trying an actual shot with a liquid-fuel rocket." He then argued that Ley's books, along with the publicity surrounding *Woman in the Moon*, "meant very little" from an engineering point of view. "The Germans," Pendray wrote, "were too busy arguing the merits of space flight to do any actual experimenting. . . . In the meantime, Goddard was going doggedly ahead, making and shooting rockets." In his obituary of Goddard, Pendray went further, stating: "On March 16, 1926, he shot the first liquid fuel rocket ever constructed . . . which was the ancestor of all liquid fuel rockets constructed since, including, of course, the German V-2 rockets." Pendray was beginning to hint that the V-2 rockets had been influenced by Goddard's designs, as if the Germans had stolen the plans. Goddard himself had believed this assertion.[24]

Ley kept quiet about Pendray's motives, for the time being. However, in late 1946 he quickly revised and expanded his own book on rockets, now titled *Rockets and Space Travel* (1947). As he revised the text, he had yet to reunite with several of his former German colleagues. That situation soon changed.

A Tense Reunion

On December 6, 1946, Ley reunited with Wernher von Braun for an evening of wine and shoptalk at Ley's home in Washington, D.C. Ley welcomed von Braun, who expressed great pleasure to see Ley after such a long absence. They enthusiastically discussed the German rocket program until 2:45 a.m. In a letter to his friend Herbert Schaeffer, Ley described the scene: "I intentionally took no notes during the conversation, so that it did not seem like an interrogation." Yet Ley memorized each point and then later recounted them to Olga. Both Ley and von Braun showed much caution with questions and answers, so that the meeting could not be misunderstood by governmental officials. Aside from learning as much as possible about the V-2, Ley learned other interesting facts. For example, a certain Colonel Riffkin had read his *Rockets*. In fact, this official often quoted the book directly, and he expressed an interest in accompanying von Braun. Ley was also surprised that von Braun knew something about his own work for WIT. Perhaps most noteworthy was Ley's judgment regarding von Braun as an engineer who worked for the Nazis: "I found no reason to regard v.B. as an outspoken anti-Nazi. But just as little, if not even less, did I find him to be a Nazi. In my opinion, the man simply wanted to build rockets. Period."[25]

On the day of the reunion Laning wrote to Heinlein: "He [Ley] is going to have to play pretty hard to get into the rocket picture. For some time only the military will finance it, and will control the contractors. Yet Willy must be keep abreast. How will he do it I don't know." When Heinlein learned of Ley's meeting with von Braun, he expressed his revulsion, telling Laning, "I find the whole matter very distasteful." Laning responded by defending Ley's actions: "I fear I can pardon him. This damned kowtowing to von Braun seems to have made it necessary for anyone, who wants to know the rocket field, to get information from that former Nazi. Willy has not had a very fair deal in this country as far as employment of his talents is concerned." In spite of this defense, Laning cautioned Heinlein by stating, "I'm sticking my neck out to vouch for him; if you have information please warn me." Later in the month Heinlein clarified that he had no reasons to doubt Ley from a political standpoint. Yet the meeting with von Braun diminished his opinion of Ley's judgment. He told Laning: "I am not willing . . . to sponsor him any longer." Ley's further attempts to secure a navy contract would be futile. He lost the support of Heinlein and Laning due to fraternizing with von Braun. He would continue to work as a research engineer until October 1948. Yet he gradually gave up hope for a contract, and he returned to a typewriter as his chosen machine.[26]

Rockets and Space Travel (1947)

In early 1947 Viking released a revised and expanded edition of *Rockets*, retitled *Rockets and Space Travel: The Future of Flight Beyond the Stratosphere*. In the foreword Ley explained the new title: "This is a book about rockets and about the idea of interplanetary travel, and I wish to emphasize that these two things belong together." He also stated, "Because it is my firm conviction that rocket research will lead to the realization of that great old dream and because I see little value in any rocket research which states that it is not supposed to lead to that goal, I have written this book." On the one hand, Ley still downplayed the future evolution of war rockets by arguing that space travel was the goal of rocket research. On the other hand, he removed the passage that claimed that the story of the war rocket "lies in the past." Instead, he presented the V-2 as "merely the beginning." Future rockets would rise higher. After that, "the spaceship will follow . . . one day in the future. Possibly in a future not too distant." He also removed a passage that stated, "The modern war rockets do not replace artillery in any way; they merely augment it."[27]

Ley made several small revisions to existing chapters. The organization of the book remained the same, apart from the inclusion of two new chapters called "The Rockets of the Second World War" and "Peenemünde!" Ley's narrative surrounding the V-2 is interesting and revealing. Mostly it indicates what he did not know about the production and technical details of the V-2. The chapter can be read as an American science writer's catalogue of known facts. Ley did not evaluate the V-2 in kind terms, although he glorified the broader implications. The V-2 was "not fully developed" and "showed a number of glaring imperfections." He added, "A 'usable state' was good enough for the hard-pressed Germans." Nevertheless, the V-2, in spite of its failure to alter the course of the Second World War, "transformed the face of war for all time to come."[28]

Ley's inside knowledge of V-2 rockets and Peenemünde became a key selling point. In retrospect, his knowledge in 1947 was minimal. However, there are a few passages that indicate last minute revisions of the text, possibly based on knowledge of von Braun. For example, Ley told this story:

During 1943 Count von Braun went to see Hitler at his headquarters at the eastern front. With him he had rolls of film, documenting the research work done. Apparently both von Braun (who happens to look like the picture of the "perfect Aryan Nordic" invented by the Nazis) and his films impressed Hitler sufficiently to make him change his mind. He ordered mass production.[29]

Otherwise Ley relied more heavily on newspaper accounts than information from the ex-Peenemünders. Yet, as historian Neufeld argued, Ley's glorification of the rocket research center was the first step toward a more "sanitized history of Nazi rocket activities palatable to Western audiences during the Cold War." Ley knew little about the atrocities "due to a deliberate policy of silence by the ex-Peenemünders and the U.S. government." Inadvertently, he crafted a narrative that served their interests.[30]

Other revisions reflected an enormous change in Ley's evaluation of the V-2 and aerial weapons. In the past he had consistently told the public to "keep calm" due to the evolving balance between offensive and defensive weaponry. Now Ley could offer no valid counterargument to the "prophets of doom." He admitted, "But all these arguments pro and con are invalid now; they have been cut short by the atomic bomb." Although the V-2 remained an inaccurate missile, "it becomes the final weapon if it carries an atomic bomb. . . . There is no defense."[31]

The book also promoted Cold War perceptions of a space race. Although Ley mostly avoided discussing the military applications of space technologies, the implications of the text were obvious. In fact, science writer Martin Gardner reflected on the book in the *Scientific Monthly*: "Now . . . it is evident that space travel is only a few years away and that the first nation to establish a military base on the moon will dominate the earth." Ley encouraged this type of thinking with extremely optimistic accounts of the pay-offs for the first nation that constructed a space station: "The station in space promises many new discoveries. It is not impossible that a single one of them will pay for everything."[32]

Some readers may have been disappointed by Ley's superficial account of the V-2, since he lacked insider knowledge. However, most reviewers praised the book, while reaffirming Ley's unique status as a foremost rocket expert. The *Field Artillery Journal* claimed, "This is *the* book on rockets and space travel." Gardner disagreed: "Ley's book should be read in conjunction with G. Edward Pendray's *The Coming Age of Rocket Power*." During these months one might suspect that Ley worked behind the scenes to rival Pendray. For example, after the *Coast*

Artillery Journal printed a Pendray extract that focused on Goddard, Ley shot back. The March–April issue contained two original articles titled "The Problem of the Step Rocket" and "The Interception of Long-Range Rockets" as well as an extract from Ley's book. Ley also allowed the *CAJ* to reprint his "Limitations of the Long-Range Missile," which originally appeared in *Ordnance* magazine.[33]

Meanwhile, Ley likely grew disturbed by the rising anti-German sentiment regarding the U.S. military's employment of the ex-Peenemünders. The presence of the rocketeers was now public knowledge. As historians have noted, the public outrage centered on the amorality of using former servants of the Third Reich, who were seen as morally compromised. This anti-German sentiment fit well with Pendray's efforts to nationalize the history of rockets by making it an American story. While critics wanted the German engineers expelled, Pendray wanted to expel the German pioneers from the historical record. Pendray's agenda and the broader anti-German backlash threatened Ley's professional and financial situation, after the birth of his second daughter, Xenia.[34]

Although Ley would later defend this group publicly, he remained rather silent during 1947. He may have had mixed feelings. On the one hand, this was a time when his anti-Nazism hardened into a more virulent anti-communism. If he had known the true extent of von Braun's activities, as well as the use of concentration camp labor in the construction of the V-2 rockets, Ley might have severed his ties to these Germans. On the other hand, he viewed them as apolitical engineers who were forced to work for the Nazis. Additionally, Ley was now dependent upon these Germans for information about the V-2 program. His reputation, as well as the livelihood of his family, depended upon his success as a freelance writer. The Peenemünders, particularly von Braun and W. Dornberger, would become indispensable sources of information.

Technology Review

While continuing to write for other publications, Ley became more directly involved with MIT's *Technology Review*. Not only would he serve as an editorial associate, but he would also contribute a dozen articles on science and technology. In many ways this move indicated Ley's ambitions to associate himself with a more respectable publication, while he worked as a research engineer. Indeed, MIT's *Technology Review* was a "who's who" of influential scientists and science writers. For example, Ley shared staff responsibilities with the president of MIT, James R. Killian Jr., along with Philip M. Morse, who is considered the "father" of operations research. Often Ley's articles appeared alongside influential writers. For example, the January 1946 issue showcased Ley's "Fortunes—Twenty Fathoms Down" alongside "Research on Minority Problems" by psychologist Kurt Lewin and "Science and the Civil War" by historian of science I. Bernard Cohen, later the editor of *Isis*. Other contributors included anthropologist and science writer M. F. Ashley Montagu, engineer Harold E. Edgerton, and mathematician Paul Cohen.[35]

Science administrator Vannevar Bush contributed frequently. Indeed, some of Bush's articles outlined the overall agenda of *Technology Review*. For example, Bush argued, "Science imposes new duties on scientists and engineers in the problem of attaining national harmony and international peace." It was not enough simply to have faith in science, because "faith without work is not enough . . . there is work, much and great work, to be done." Scientists had an "ethical imperative," with "two great ends" of influencing American public opinion and fostering world enlightenment. Scientific truths were international truths. Therefore all scientists spoke a common language in their "ministry to the people." By broadcasting their message, they would spread the gospel. In a later article Bush expanded on "the scientific way" that encouraged investigation, rationality, and free thinking.[36]

Technology Review preached the gospel. For example, it printed a speech by J. Robert Oppenheimer, who argued, "Of all intellectual activity, science alone has . . . turned out to have the kind of universality among men which the times require." This universality could be seen most directly in the West, whereas science was in a general state of "decay" in Soviet zones because politics and terror had "corrupted its very foundation." Oppenheimer expressed an enormous amount of optimism about science in a free and democratic society. Other writers focused on the potential of science and technology to bring international peace.[37]

Various articles included pleas for scientific education as a means to train citizens to think rationally and reject the totalitarian mentality. In particular the magazine published James R. Killian Jr.'s valedictory address, in which he argued that "tolerance, willingness to accept change, and faith in the future and in our spiritual unity stand as bulwarks against doctrines of regimentation." Against the threat of communism stood "knowledge about our American democracy and its special ideals and values." This knowledge was "the strongest weapon" against "the zealot." Through science, Americans possessed "a tolerance of differences, a repugnance for regimentation, and an acceptance of dissent" as well as "the spiritual unity of our people." The speech concluded with a dire warning: "You have a mandate . . . to live the American credo if you want the American credo to live." A different article by physicist Arthur Compton demanded "Education for Peace." He argued: "If this democratic society is to compete successfully with dictatorial civilization, it must likewise develop leaders of outstanding professional competence, dedicated to honorable action in the interest of human welfare." It was equally essential to train ordinary Americans: "Our strength lies in the many millions of our citizens who are working efficiently and loyally at the nation's tasks." The speech blended well with other articles, which stressed the role of education. It was imperative for scientific intellectuals to gain large audiences. They should write popular books. Joseph H. Keenan proclaimed: "Nothing less than a wide dissemination and a general acceptance of the principles of free inquiry and of individual responsibility will do." Educators must "exemplify the spirit of free inquiry and the sense of responsibility." They must inspire students to engage in the same type of intellectual freedom.[38]

The broader mission of MIT fit well with this Cold War rhetoric. For example, in "President Killian's Statement of Academic Freedom and Communism," Killian argued: "The Institution is unequivocally opposed to Communism; it is also sternly opposed to the Communistic method of dictating to scholars the opinions they must have and the doctrines they must teach." MIT's researchers "must be free to inquire, to challenge . . . to doubt . . . to examine controversial manners, to reach conclusions of their own, to criticize and be criticized." Yet a line could never be crossed. "The teacher, as a teacher," he argued, "must be free of doctrinaire control originating outside of his own mind . . . and above all he must work in the clear daylight without hidden allegiances or obligations which require him to distort his research or teaching in accord with dictates from without."[39]

Outside the pages of *Technology Review*, Vannevar Bush took this anti-communism further in his book *Modern Arms and Free Men* (1949). The text illustrates how these views on science transitioned into critiques of the totalitarian alternative. Bush argued, "The weakness of the Communist state resides in its rigidity to the fact that it cannot tolerate heresy, and in the fact that it cannot allow its iron curtain to be fully penetrated. All these things, vital to totalitarianism whether right or left, are fatal to true progress in fundamental science." Bush added: "Dictatorship can tolerate no real independence of thought and expression." Scientist/historian Conway Zirkle made similar points in his 1949 book *The Death of a Science in Russia*. A reviewer remarked, "History proves that science cannot thrive under totalitarian control of any kind. Not since 1633 . . . has science been so threatened by authoritarian control of the human mind." Like Galileo's inquisition, the scientist stood on trial.[40]

A scientific democracy could never tolerate the authoritarian's prejudiced and dogmatic thinking. Instead, it became crucial to correct the authoritarian's distortions of reality. According to sociologist George A. Lundberg, science could serve as "a sort of mental hygiene." It would provide a "unified method of attack." Eventually it would triumph over "magical thinking" and "prescientific modes of thought." "All that needs to be deplored," Lundberg added, "is the [layman's] inability to distinguish between fact and fable, the practical and the fantastic." Lundberg concluded: "When we give our undivided faith to science, we shall possess a faith more worthy of allegiance than many we vainly have followed in the past."[41]

For this broader camp of intellectuals, the history of science became useful, in terms of educating and uplifting the masses. Most famously, chemist and Harvard president James Conant outlined an agenda in his 1946 lectures, published as *On Understanding Science*. Conant praised the scientists, yet held special praise for the public, which could embrace scientific thinking. For Conant, historical case studies could teach non-scientists how to see through the eyes of a scientist. Thus the history of science became a key tool to educate the masses and debunk irrational beliefs. The lessons of the past would help to create "a unified, coherent culture suitable for our American democracy in this new age of machines and experts." In this sense, using the history of science to train citizens was akin to using military history to train soldiers. Chemical engineer Thomas S. Sherwood likewise argued, "A double purpose of our educational system might well be to acquaint the citizen with the true meaning of science, while broadening the training of scientists and engineers in the humanities." Scientists should do more than research. They needed to speak out, in intelligible language. They needed to reach the masses.[42]

Willy Ley contributed to this scene by writing about the past. He sought to unite science and the humanities. His histories of science included a moral crusade. He fulfilled Bush's hopes that the history of science and technology "are of profound worth in this search" for the truth. "The past," Ley wrote, "proclaims the future."[43]

Another Excursion into Romantic Zoology

When a distinguished professor spoke at MIT's "Alumni Day 1948," he announced: "The history of living things . . . teaches something more inspiring than what we are often inclined to take with us from the teachings of the physical sciences. . . . There is something extraordinarily encouraging and interesting in these facts of biology." Ley shared the professor's excitement for the history of living things. He loved to explore the mysteries, untangle the facts, and recall the past adventures of scientists. He also enjoyed sharing his love of exploration, by writing educational and entertaining articles for magazines. As his tenure as a research engineer came to a close, Ley wrote dozens of articles and several new books. Many of his works read like detective stories about animals, myths, and legends.[44]

Notably, Ley produced a revised edition of *Lungfish*, now titled *The Lungfish, the Dodo, and the Unicorn* (1948). The themes of this book remained intact. What is noteworthy about the new edition can be seen in its public reception. Whereas the first edition had largely failed to attract a broader audience, the 1948 edition generated sales and publicity. Vincent Starrett of the *Chicago Daily Tribune* called it "one of the most fascinating books of our time." A different reviewer described the book as "an astonishing zoological garden." It also showed how "truth, as everybody knows, is sometimes stranger than fiction." Consequently, this "most enchanting of recent books . . . [was] irresistibly readable from the first page to the last." Orville Prescott of the *New York Times* praised the book's "craving for the miraculous." Overall, he wrote, "it is an unusually able and interesting example of popularization of science."[45]

W. M. Mann of the National Zoological Park also praised the book for both its scholarly and entertainment value. He wrote, "Willy Ley . . . has gone through an enormous amount of classical and medieval writing, and assembled his clues with as much suspense and thrill as would the writer of a modern 'who-dun-it.'" The book as a whole was "so delightful and well done." Other science writers and historians agreed. In the pages of *Isis*, geneticist-turned-historian Conway Zirkle wrote: "Perhaps the basic moral to be drawn is that nature can equal art even when art is most imaginative, and that science can match legend in romantic interest." None other than *Isis* founder and Harvard historian of science George Sarton praised the book. While nursing a cold in 1951, he read the new edition. He then wrote a personal letter to Ley, saying, 'I . . . enjoyed it so much that I feel moved to express my thanks.'" The letter ended with the words, "Bravo! And best wishes to you."[46]

Ley's 1948 edition of *Lungfish* was his most popular success to date in the history of science. It explored the mysteries of nature. It glorified the unknown. It celebrated the brave explorers and their daring adventures. It led readers to see the world through the eyes of the explorer.

Readers could embrace scientific thinking without losing an iota of thrill, wonder, and enchantment. Ley trained his audience to embrace the scientific way.

The Conquest of Space (1949)

By October 1948 Burke Aircraft Corporation's links to the WIT were terminated. Ley no longer had an active position as a research engineer, although his contract did not expire until March 1, 1949. His brief tenure as a consultant for the Department of Commerce soon ended. His cumulative experience as research engineer involved a game of endless negotiation, contract bids, and fights with machinists. Most likely the situation grew worse following the budget cutbacks of 1947 as well as the slow developments and departmental competition that frustrated von Braun. In general, Ley felt discouraged by the lack of progress. Apparently his books, and the combined efforts of other science writers, had produced very few results. Throughout much of 1949 Ley wrote fewer articles on rockets and spaceflight. There are some exceptions, such as his piece for the *Rotarian* called "Want a Trip to the Moon?" After Burke and WIT parted ways, the Leys moved from Washington, D.C., to Montvale, New Jersey. They moved into the house of Olga's mother, Dr. Maria Feldman. According to Ley, it was a "beautiful and large house on a hill, near a forest." Ley would spend two years in this rural setting, isolated from the conveniences of a big city, which Ley preferred. In a letter to Heinlein, Ley described the "frightfully idyllic and equally boring" setting.[47]

Meanwhile, a Hollywood artist gained momentum for the cause of spaceflight. Chesley Bonestell had become an accomplished backdrop artist for films such as *Citizen Kane* (1941). Having an interest in astronomy, he also painted alien landscapes, which he submitted for magazines, such as *Life* in 1944. By the late 1940s Bonestell had become quite wealthy, due to his paintings and astronomical knowledge. Co-authors Ron Miller and Frederick Durant explained: "He was literally teaching himself astronomical painting." Bonestell's interest in astronomy led him to consult with experts at Mount Wilson observatory and the Hayden Planetarium. One expert introduced Bonestell to Ley.[48]

Bonestell and Ley produced *The Conquest of Space* (1949), a bestselling coffee-table book. The publication was a watershed moment for the cause of spaceflight. By 1950 the book achieved its fourth printing, before reaching an international market, from Italy to Japan. That international audience soon expanded due to Dutch, Finnish, German, French, Swedish, and Spanish editions. It is easy to see why the book was successful. Historians have rightly pointed to the beautiful paintings by Bonestell, which depicted the earth from space while giving readers a close-up view of the moon, Mars, and even the distant planets of our solar system. Historian Howard McCurdy is right to showcase the paintings as the book's "most distinguishing features," while Bonestell's name received top billing. The images were stunning, not only for their scientific accuracy but also for their imaginative insight. Many of Bonestell's illustrations resembled black and white photographs, meant to represent the first images of a trip from the earth to the moon. Bonestell painted them in color, but the black and white prints had a more dramatic effect. Not only do readers experience a first person glimpse of the journey to the moon; they also witness the first transmissions from spacecraft. Bonestell's paintings, according to Ley, were "the product of a poetical mathematician with a paint brush." Each image served as "a picture."[49]

Historians often focus on the futuristic and prophetic elements of the book. Ley's text complemented the prophecy, as the first chapter dramatically described a rocket launch. Ley placed readers on the "innumerable sand hillocks" of the White Sands Proving Ground, as they anticipate the launch under a hot sun. The text then described the dramatic launch of an unmanned V-2 rocket: "'Six'—'Five'—'Four'—now the turbine is running—'Three'—'Two'—now the turbine-driven pumps *force* the fuels—'One'—the noise of the rocket has become incredible, deafening; impossible sound wave piled on impossible sound wave—'Zero!'—'Rocket away!'" As the rocket rises, the spectators marvel. The scene then shifts to a classroom, where a professor teaches his students about mathematics and rocket trajectories.

While encouraging them to "leave all earthbound concepts behind," he brings in concepts from astronomical science. The lesson concludes with predictions regarding the scientific uses of an unmanned orbital rocket and a space station.[50]

Conquest moves into chapter 2, called "Target for Tonight: Luna!" Given the prominence of the moon in the history of astronomy, it is a "small wonder" for Ley "that all the speculations, thoughts, and dreams which we are now tempted to label the 'prehistory of space travel,' concerned the moon and the moon only." Ley then explained the physics of an unmanned trip to the moon. The moon rocket is technologically feasible and "nearly within reach of present day technology." Ley conceded: "There would also be a lot of new problems of all kinds, leading to torn sketches and torn hair in the engineering department. . . . In general, however . . . its design and construction are possible without any *major* inventions. Its realization is essentially a question of hard work and money."[51]

The manned rocket to the moon, on the other hand, "is a different story . . . beyond our present ability." Although "the 'how' is still to be discovered," it is easy to imagine such a voyage: "We know that it will begin with tense minutes of waiting on a mountain top near the equator. . . . We know that finally there will be zero hour, zero minute, and zero second." Ley then takes readers on the journey to the moon, while Bonestell's illustrations grow closer and closer to the lunar mountains. Readers are encouraged to imagine this dramatic moment of exploration and discovery, as the bold pilot penetrates a new realm of nature. Ley described the incredible feat: "The earth will be a monstrous ball somewhere behind the ship, and the pilot will find himself surrounded by space. Black space, strewn all over with the countless jewels of distant suns, the stars." The third great era of astronomy was now officially underway. Whereas the first great era involved the naked eye and the second great era surrounded the use of instruments on earth, this third era ushered in the actual conquest of space. Ley spends much of the remainder of the book discussing the history of astronomical theories, while often arguing, "We'll never know until we get there."[52]

The Conquest of Space was an instant hit for Ley. By February 1950 it had sold almost 20,000 copies. It was also well received. One reviewer proclaimed, "Hold your hats, every one! We're off to the moon. . . . So, hold tight! Here we go!" John E. Pfeiffer reviewed the book for the *New York Times*. "This book," he wrote, "is the latest and, in many respects, the most fascinating popular account of rocket travel and what people may see when they reach various landing places in the solar system." He also praised the artwork of Bonestell: "The combination of his picture gallery and Mr. Ley's text makes this one of the year's best popularizations of science." Few reviewers critiqued the combination of a popular science writer and a Hollywood artist. According to the *New York Times Book Review*, "It would be difficult to find two men better qualified."[53]

Many reviewers judged the book to be scientific and serious, despite its coffee-table format. According to the *Nation*, the book was "responsible fantasy." A reviewer for the *Scientific Monthly* even compared it to Renaissance classics: "In the early days . . . a scientific treatise was not only informative but also a work of art. The revival of this old custom has been successfully achieved in this book." The book combined Ley's "lucid discussions" of astronomy and spaceflight with Bonestell's work as an "architect, astronomer, and artist." Lastly, this "outstanding Viking publication" would appeal to both general readers and "the technical student who wishes to have a reference book he can enjoy." Other reviewers praised the book widely. Not surprisingly, Robert A. Heinlein wrote a lengthy article for the *Saturday Review*. "Until the first rocket lands on the moon," he announced, "this book is the next best thing to interplanetary flight."[54]

By March 1950 the success of *Conquest* convinced the Hayden Planetarium in New York City to "run a show based on the book." Ley informed Heinlein, "We don't get anything for this directly, but they are going to sell the book in the lobby of the Planetarium which should help." Heinlein congratulated Ley: "More power to you! I hope that the sales keep up forever, and that you become inordinately rich. It's a fine book and I never miss a chance to plug it."[55]

It would not be long before Ley could afford to move the family back to New York City. Ley probably viewed the royalties as a small bonus. The real success of his book remained to be seen. It would be up to a younger generation to embrace the cause. He wrote to von Braun:

"The old guard is too hard-boiled to stop preaching space travel and the younger generation will be, I hope, sufficiently impressed to go on. Some of the explorers of the Arctic have written in their memoirs that they embarked on their career because when they were boys they saw pictures of arctic landscapes." A new generation would dream of making similar journeys after studying the illustrations of Bonestell. The images would fuel their excitement. Imagination would soar to new heights. Reality would be molded to fulfill human expectations. Engineers' dreams would influence actions, if the momentum could be sustained. Ley positioned himself to become the publicist for the cause.[56]

The Tom Corbett Years

In June 1955 Willy Ley went to NBC studios in New York City. They had a special job that required his expertise. Ley would serve as the brain of TV's space cadet, Tom Corbett, during a public appearance. As actor Frankie Thomas Jr. answered the questions of a large group of fans, Ley stood behind a curtain, communicating to the actor via a hidden microphone. A journalist described the scene: "If you think conditions around Mars or Saturn are tough for Tom Corbett, you ought to see what headaches he runs into when he comes back to earth!" He continued: "Hardly had the imaginary jets of his rocket 'Polaris' cooled off in the NBC studios when Mr. Corbett is pelted with the goldangdest bunch of questions ever dreamed up by the wizard minds of 20th century boy and girl." His fans asked about rockets, missiles, and space travel. With Ley speaking softly in his ear, the actor answered the questions. After the ordeal, the actor reflected, "I might be okay out in space, but I'd sure be lost in those quizzes without Willy. . . . You can't bluff your way through. . . . They ask questions like, 'How can you control direction in a vacuum'—and if you can't answer, you're a 'jethead.'" The actor added, "I thank my lucky stars that Mr. Ley is my co-pilot."[1]

By 1955 Ley had done so much to become the man behind the curtain of the Space Age. His *Conquest of Space* had become a bestseller. During the early 1950s new editions of his *Rockets* provided authoritative guides. While science fiction authors used this book as a technical bible, Ley served as a consultant for television, books, and even a comic strip. Thus many of the images of rockets, space stations, and men in space were influenced by his books, articles, and public lectures. Ley became an influential publicist and popularizer. He persuaded millions of Americans to believe in space travel. This success was a part of his broader accomplishments as a science writer. He crusaded to educate and entertain, while relentlessly evangelizing for the scientific spirit. By viewing his space-related media as part of a larger agenda, we can situate his work within a broader scene of postwar optimism, celebratory media, and experiential sites. Ley's influence can teach us much about rising public confidence in the years prior to *Sputnik*. Along with von Braun and other influential figures, he promoted an inevitable future of Americans in space.

Destination Moon

When *The Conquest of Space* hit bookstores, momentum for the cause was increasing. The most encouraging sign, besides the sales of books, was the news of a major Hollywood movie, to be produced by George Pal and Paramount Pictures. It was titled *Destination Moon*. By December 1949 commentators spoke of a space race between scientists and moviemakers. It was obvious who held the advantage. A *Los Angeles Times* reporter proclaimed: "It looks as though Hollywood will get there first—and in Technicolor." In some ways the film was similar to Ley and Bonestell's *The Conquest of Space*. Bonestell was directly involved with the production of the film, which included astronomical backdrops. The film also benefited from the combined efforts of Bonestell and a "No. 1 technical adviser," Robert Heinlein. As the chief consultant, Heinlein strove for scientific accuracy. He used Ley's *Rockets* as a reference text.[2]

Ley had little to do with this production. His lack of involvement almost ruined his friendship with Heinlein. In June 1949 Heinlein sent a postcard with technical questions. Ley responded coolly, "I feel that a movie company ought to pay for information if it wants it . . . tell them please that I have two children to support, that my time is my

merchandise and that it is, therefore, for sale." Ley felt hurt that Heinlein did little to include him in the production. Ley may have viewed
the situation as a betrayal. When Heinlein received Ley's response,
he immediately wrote a long and apologetic letter. Heinlein claimed
to have campaigned for Ley as the film's consultant. Unfortunately, a
contract was impossible due to the film's budget. Heinlein added his
reason for sending the question: "I guess I was too goddam [sic] subtle,
but the purpose of that question was simply to 'invite you into the
club,' let you know that you had not been forgotten." Ley responded
two months later, after Heinlein wrote a generous review of Conquest.
Ley corrected Heinlein's version of events: "You hadn't said anything
before, nobody else knew anything. . . . If you had made it clear from
the outset that you were bringing personal sacrifices my reaction would
have been personal too,—instead of the attitude of self-defense. . . . So
that dog is daid [sic]." Ley would have other opportunities to influence
a mass audience.[3]

Tom Corbett, Space Cadet

In October 1950 a TV show premiered on CBS. It was the first episode
of one of the longest running television programs of the early 1950s.
The show began: "Kellogg's, the greatest name in cereals, presents
Tom Corbett, Space Cadet." An image of a V-2 rocket cut to a control
room, as the narrator announced: "This is the age of the conquest of
space. 2350 A.D: The world beyond tomorrow. . . . In roaring rockets,
the space cadets blast through the millions of miles from Earth to the
far-flung stars, to protect the liberties of the planet, defend the freedom of space, and safeguard universal peace." In the first scene, space
cadets chanted, "To this end, I dedicate my life." Then an incoming
rocket pilot demands to speak with unit leader Captain Strong, because the "safety of Earth depends on it." Viewers meet cadets Tom
Corbett, Astro, and Roger Manning, as they attend a lecture. Captain
Strong begins:

> You, as Space Cadets, will now start training to become officers of
> the Solar Guard. Your responsibilities are great, men. You hold the

future of the Solar Alliance in your hands. . . . There are people to be met and understood, but remember this, and remember it well . . . you'll meet them as men of peace. You'll deal with them in honor and trust. You'll fight only for freedom and for liberty.[4]

Space Port Control interrupts the lecture to summon Captain Strong. Astro and Tom grow concerned. In the next scene, Strong assists the incoming rocket that has yet to "turn tail." The captain of the ship is disoriented. Meanwhile, Astro introduces Tom to the "rocket-cruiser of the latest design" called the *Polaris*. "It's beautiful," Tom exclaims, "the most beautiful thing in the whole world." Astro delivers a monologue:

Aye . . . and one day if you're lucky, you'll be the master of such a ship as this. You will walk her decks a million miles out in the void, stare out of the view ports into the majestic blackness of outer space. You will see planets, asteroids . . . worlds dead and worlds still unborn. Then, by all the satellites of Jupiter. . . . Then, Tom, you will know what it means to be a space man![5]

"You make it sound pretty wonderful," Tom responds. Suddenly Astro and Tom notice the incoming rocket. Captain Strong rushes to the boys, forcing them to the ground. They witness an explosion. "It's all over . . . the end for them," Astro laments. Strong responds, "For them, yes. I'm afraid it's only the beginning . . . for us."

Airing just before prime time, three times a week, *Tom Corbett, Space Cadet* narrated the daring twenty-fourth-century adventures of a young man and his small crew, as they journeyed through the solar system. As cadets of "Space Academy, U.S.A." (later changed to "Atom City"), the crew rocketed into the cosmos, where they enforced the laws of the Solar Guard. Often the central conflict involved the clashing personalities of the main characters. Comic relief came in the form of space slang, such as "Cool your jets!" and "Blast me for a Martian Mouse!" *Tom Corbett* was not the first science fiction program to appear on television. It followed in the footsteps of *Captain Video* and *Space Patrol*. However, the show distinguished itself as the most plausible. *Newsweek* commented, "Space Patrol and Captain Video shows sometimes extend action beyond the possible, but the Tom Corbett program makes a point of keeping actions within the limits of scientific accuracy." Although early scripts contained some wild ideas, "Ley will have straightened out most of them." With Ley as the technical consultant, the show "generally provides its audience with possible—though still unrealized—facts, and juvenile watchers are getting science lessons along with their entertainment." *Newsweek* concluded, "If the moon is experimentally reached by man-carrying rockets in 25 years, as Ley predicts, it will be rather old stuff to many of today's youngsters."[6]

Ley often gave the impression that his work for *Tom Corbett* was an insignificant sideline. For example, in a 1957 interview, journalist Mike Wallace asked: "Does it ever bother you that you're also known as a consultant for the science fiction TV series *Tom Corbett, Space Cadet*, and . . . as a special advisor to the Sugar Jets cereals' space commercials on ABC's *The Mickey Mouse Club*?" Without missing a beat, Ley responded, "Well, no that doesn't bother me at all, because if I weren't there, I feel somebody else might do it and talk nonsense. So, I almost feel like an educator by bringing in things which I consider correct." In personal correspondence, Ley expressed hope for the show. He told Heinlein, "Mind you, I don't claim that the show is perfect as of now. But it has much improved and there is hope that it will be something to be proud of in a few months." At a time when many literary critics damned television and its effects on children, Ley embraced the potential of the medium both to educate and to entertain. Like Don Herbert (whose *Watch Mr. Wizard* appeared in 1951), Ley perceived an enormous opportunity to bring science into the homes of American families.[7]

Lead actor and co-writer Frankie Thomas Jr. recalled, "When I was writing some of the *Corbett* shows . . . I would have a conference with Willy, and it was amazing, the things that he said *we could do*, things which he said *were* in the realm of scientific possibility." Ley generally approved most of the writers' ideas. Thomas remembered, "He would check your story's 'scientific possibility'—those were the words he used—and if it was scientifically possible, okay, go ahead with the story!" He argued, "We didn't go for horror stuff. We tried to stick with 'the scientifically possible.' We were totally different from *Space Patrol*. They had monster shows and Dracula-like characters. We had man-against-man a little, man-against-nature and man-against-him-self." Thomas added, "All our stories were run by Willy Ley, and he would make sure that the stories involved things that were scientifi-cally possible. They didn't have to exist, but they *could* exist." Ley also took pride in this work, telling Heinlein, "Fortunately most of the people involved got their first 'familiarization' by reading my books, so they listen when I open my mouth. . . . I veto scientific impossibili-ties." His biggest complaint surrounded the setting of the show in the twenty-fourth century. This future would unfold much sooner.[8]

Additionally, Ley influenced the show's publicity campaign, after producers sent the cast to Pittsburgh, Philadelphia, and Boston. These public appearances generated crowds of young fans who sought au-tographs and toys. In a later interview Thomas remembered a par-ticular appearance in Philadelphia: "There was a line all the way out the door—there were 10,000 people there! I was shaking hands and saying, 'Spaceman's luck!' and all of that . . . there wasn't *time* to sign autographs." While this estimate is likely exaggerated, the show's mar-keting campaign was extravagant. It attempted to re-create the craze of *Hopalong Cassidy*, a fictional cowboy hero on radio and television. Pro-ducers hoped that children would exchange their cowboy hats and toy revolvers for space suits and decoder rings. Meanwhile, Kellogg's ce-real offered *Tom Corbett* toys and prizes. When Ley discovered that an apparel convention in New York City included the promotion of works by Bonestell, Heinlein, and himself, he suggested that they amp up the marketing campaign. Ley described his involvement: "I said: why not Bonestell originals? Why not a few actual rockets? So now there is a combine consisting of the Tom Corbett program, the Viking Press and the Hayden Planetarium (I brought them all together) working on a permanent exhibit, designed to travel." The exhibit mixed science and entertainment. Ley remained a key organizer, bringing experts, publishers, and celebrities together.[9]

Rockets, Missiles, and Space Travel (1951)

As he organized the conventions Ley revised a new edition of *Rockets*, now titled *Rockets, Missiles, and Space Travel* (1951). He told a reporter,

"The new book is essentially a history of the development of rockets from the beginning to the future, to the moonship, to landing on the moon and building a base there. . . . It is a history up to the present. After that it is prophecy." Because Ley now relied on the ex-Peenemünders as sources, one might expect the book to match historian Neufeld's description of "a romanticization of the Nazi rocket center . . . as fundamentally aimed at space travel, rather than weapons development for Hitler." This romanticization creeps into the text. Ley described the initial site as "strung along the seashore, with laboratories, workshops, test stands, etc." Ley even asked if the site was "a research engineer's paradise," though the engineers were "operating for the wrong cause." Ley's book also included dramatic countdowns of V-2 launch tests, which conflated the V-2 with a spaceship. Nevertheless, Ley identified the V-2 rocket as a weapon of war, commissioned for only one purpose: as a missile.[10]

By this point Ley's *Rockets* had grown to 436 pages. His agenda had changed. Whereas the 1944 edition tried to persuade the public "that he was serious," Ley asserted, "The question now is simply how soon engineering practice will catch up with existing theory." Due to progress in rocket research, the book became far more complicated and bogged down with equations and diagrams. Ley wrote for both laymen and specialists. He struggled to balance a readable text with technical diagrams and appendixes. Consequently many reviewers saw the book as more imposing than earlier editions. One reviewer grumbled, "He writes as a rocket scientist, with a kind of mathematical fury which may baffle well meaning laymen who never got beyond trigonometry." Yet, this critic admitted, "Mr. Ley can be technical, but he has an appealing sense of wonder and a wonderful sense of curiosity." Despite critical reviews, the book sold well. The Natural History Book Club gifted it to new members. They announced, "Here is the story of the rocket from its beginning. And here is a simplified account of present-day developments along with the thrilling story of the triumph over space that is soon to come." Ley was a prophet of the Space Age.[11]

A Scientific Celebrity on Tour

While promoting the new edition of *Rockets*, Ley went on tour as a scientific celebrity. In clubs, libraries, school auditoriums, and town halls, he talked about the past, present, and future. Ley spoke at Air Force research centers, such as Langley Field, where he addressed engineers from nearby shipyards and bases. Common lecture themes included "New Horizons—Conquest of Space" and "The Present Status of Space Rockets." He also appeared on television and radio. Ley participated in various TV youth forums in New York City. For example, he appeared with science fiction insiders Hal Clement and Groff Conklin at the *New York Times* Youth Forum. He also participated in radio conferences, such as the "WGY Science Forum." Often he accepted local invitations to speak at high schools or other small venues. His lectures and publicity activities served as a double-edged sword. He informed Heinlein, "I'm lingering a lot in front of TV cameras and other things like that. But this, unfortunately, is not a direct indicator of income. I don't complain, but most of that stuff is merely publicity and does not bring anything in directly . . . these 'honors and diversions' prevent me from writing." Ley related his situation to an unfortunate cycle, in which more time spent writing generated a demand for public appearances, yet the appearances diminished his output.[12]

Ley continued to write educational and entertaining articles. Most notable was his cover story for the *Philadelphia Inquirer Magazine*. In "You'll Live to See a Spaceship," Ley predicted, "I think it is fairly safe to say that more than half of the readers of this article will live to see a spaceship." He then argued, "As rocket technology stands right now, we could, with the fuels at our disposal, send an unmanned rocket to the moon to crash there, making a mark which could be observed from the earth through a telescope." He added, "This is as far as we can go with chemical fuels. For a spaceship [and lunar landing] . . . something else has to be called to our aid: Atomic energy." Ley predicted a seamless application of atomic power to rocketry, before claiming, "Thinking about this now is like thinking about transatlantic air flights in 1914." In an interview Ley made similar statements, arguing that "the application of atomic energy to rocket propulsion . . . is the only thing that stands between us in 1951 and actual space travel." If the imagination of Americans could be stoked, then a spacefaring future would be inevitable.[13]

Dragons in Amber (1951)

Amid lecture tours Ley found time to write a sequel to *Lungfish*. He titled it *Dragons in Amber: Further Adventures of a Romantic Naturalist*. The title had two meanings. On the one hand, it simply referenced the fossilized remains of prehistoric reptiles, which could be found inside pieces of amber. These types of "records in stone" are documented throughout the first part of the book. On the other hand, the title is much broader, because it refers to hidden gems of knowledge. These stories of interesting animals, places, and eras represented amazing case studies in the history of science.[14]

For the same reason that he wrote *Lungfish*, he wrote *Dragons* "because nobody else had." Ley elaborated: "I am surprised by one thing. Natural history is generally regarded as a rather static science which had its heyday and its revolutions during the nineteenth century. For a static science, a lot has happened to it during the two and a half decades since I sat in college lecture halls." He expressed astonishment at the fact that scientists viewed natural history as static, while they simultaneously made authoritative statements that were baseless or unimaginative. Ley debunked their proclamations and predications, while celebrating recent discoveries and revolutions.[15]

Ley then traces the history of a science back to its earliest investigators. Regarding the history of amber, he documents the early theories of Greek scientists, while referencing passages from Homer. He then documents the long struggle of a science out of medieval darkness. Later explorers had to overcome prejudices as well as misguided attempts to attribute fossils to the bones of "sinners who had drowned in the Flood." In this struggle against superstition, "There came a time, of course, when the big bones dug from the ground no longer frightened people." Science progressed most in areas that allowed for increased skepticism, the open-minded evaluation of evidence, and the international exchange of information and specimens. Ley offered readers both a celebration of internationalism and a glorification of the Western world. With politics and religion out of the way, "the scientist could no longer be restrained." Ley investigated other developments, while debunking "a lot of nonsense."[16]

Reviewers loved *Dragons in Amber*. Literary critic Orville Prescott stated, "Readers . . . know with what zestful enthusiasm Mr. Ley can write about the wonders of the natural world." Prescott added, "He has discovered that a popular touch doesn't depend on heavy breathing and high-pressure writing." He concluded: "It is all marvelously educational and Mr. Ley is a good teacher." In the *Chicago Tribune*, literary naturalist August Derleth called it "science at its best." Not only did Ley make science readable for non-specialists—he explored avenues "all too often neglected by literate scientists." Ley's persuasion, charm, and knowledge transcended "most of the best scientific writing for the general reader available today." None other than famous biologist Julian Huxley proclaimed, "The book as a whole is excellent, and will agreeably introduce its readers (including professional biologists) to many exciting facts and fascinating ideas." Huxley added that Ley was "a master in a rather unusual technique of popular science." Ley made interdisciplinary connections that specialists missed.[17]

Other reviewers commented on Ley's adventures as a romantic naturalist. For example, literary editor Joseph Henry Jackson wrote, "Mr. Ley describes himself as a 'romantic naturalist' and perhaps, in the pleasure he takes in projecting himself into the past, he is one." There was nothing contradictory about Ley's scientific mind and his "romantic temperament." Jackson stated: "In this book, Mr. Ley . . . disports himself accurately, scientifically, and in the manner of a man having a very good time indeed." Consequently Ley made science fun for the lay reader: "Mr. Ley has what amounts to real genius for tapping this native curiosity in his readers, and for making them like it." Similarly, biologist C. P. Swanson wrote, "Biology is fortunate in having so gifted a pen at its disposal."[18]

A New York Symposium

While Ley continued to excite readers about wonders, he produced another type of show. He organized the first of three conferences on space travel, held at the Hayden Planetarium. Ley recalled the origins of the idea: "One day, a nice spring day in 1951, I had lunch with Robert Coles, then the chairman of the Hayden Planetarium in New York." The discussion turned to the many "astronautical congresses which were then just starting in Europe, international meetings of rocket experts and space-travel enthusiasts." Ley lamented the low American attendance, due to travel costs. He stated, "Maybe, I said, there should be an American equivalent to these congresses." He dismissed the potential of the annual meetings of the American Rocket Society, organized by Pendray. Sensing an opportunity, Coles said, "Go ahead, the planetarium is yours."[19]

The "First Annual Symposium on Space Travel" began on Columbus Day, 1951. As an exclusive, invitation-only event, the symposium brought specialists and journalists together. Not only did Ley send invitations to institutions, societies, and research centers, but he also contacted the science editors of newspapers and magazines. The journalists became the most important attendees, due to the symposium's promotional goals. Ley argued, "The time is now ripe to make the public realize that the problem of space travel is to be regarded as a serious branch of science and engineering." If the specialists could communicate their exciting ideas, then the journalists would excite the broader public. Dr. Albert E. Parr of the American Museum of Natural History gave the opening remarks. Chairman Coles presented an imaginary tour of Mars, as seen through one of the Hayden Planetarium's "Conquest of Space" exhibits. Then Ley addressed the crowd with "Thirty Years of Space Travel Research." He reflected on past achievements in rocketry, while predicting that rockets could soon orbit the earth and beyond. Eventually, he argued, nations would construct stations in space.[20]

This symposium had a minimal impact on the general public, yet it set in motion a chain of events. The symposium intrigued the editorial staff of *Collier's* magazine, which had a circulation of 3.1 million readers. *Collier's* sent associate editor (and notable journalist) Cornelius Ryan to New Mexico, where he attended a less publicized symposium on "space medicine." Although Ley did not attend this symposium, it attracted several well-known experts like Wernher von Braun and physicist Joseph Kaplan. It also attracted astronomer Fred Whipple, who later recounted the events that took place behind the scenes. Ryan attended the talks, though he remained skeptical and often confused by the technicalities. Yet he spent a long evening of cocktails with von Braun, Whipple, and Kaplan. "The three of us," Whipple recalled, "worked hard at proselytizing Ryan and finally by midnight he was sold on the space program." Von Braun, in particular, displayed his talents as "one of the best salesmen of the twentieth century." What follows has often been called a "watershed moment" in the history of spaceflight.[21]

Collier's Magazine

After Ryan returned to New York he approached Ley and other space-flight advocates. This coordination led to the first installment of an eight-part series of articles that promoted and lavishly illustrated the cause of space travel. The first issue was dated March 22, 1952. The cover displayed a beautiful Bonestell rendering of von Braun's design for a winged rocket. The headline read, "Man Will Conquer Space Soon: Top Scientists Tell How in 15 Startling Pages." Von Braun, Ley, Whipple, and Haber contributed articles that projected optimism about the future. A second issue focused on the "next step" from the space station to surface of the moon. The editors predicted, "We will go to the moon in the next 25 years." Von Braun explained the technicalities, while Ley described the passenger rockets.[22]

Six more space-related issues appeared throughout 1953 and 1954. They became more focused on the selection, training, and health of future astronauts. They also predicted the next steps in cosmic exploration, which culminated in von Braun's plans for a voyage to Mars. By this point Ley became less involved with the *Collier's* issues. He did contribute to Ryan's expanded collections *Across the Space Frontier* (1952) and *The Conquest of the Moon* (1953). Nevertheless, by 1952 Ley had other passions that grew increasingly important. Ryan and von Braun grew to dominate the series. In many ways *Collier's* reflects von Braun's entry into the media arena. Although it was a crucial moment in the association of von Braun with the cause, along with the support of an influential magazine, the importance of *Collier's* can be overstated. For example, Bob Ward argued that von Braun's work for *Collier's* was "the first major breakthrough in spreading the space gospel to the American people." Accordingly, the series "created a sensation." It is far more accurate to agree with historian Neufeld. The *Collier's* series was an influential milestone, and it marked an important moment of von Braun's appearance on the scene of popularization. However, if viewed from a much larger context of popular science, film, television, and print media, then *Collier's* was following a trend.[23]

There may be a different way to view these issues, due to the emerging tensions of the scene. These issues stirred controversy, not only among the general public but also within the growing community of advocates. By the time that Ley organized the second symposium in late 1952, tensions flared between von Braun's unrestrained optimism and the more down-to-earth approach of rocket engineer Milton W. Rosen. Ley served as a mediator at times, according to Rosen. He later recalled, "Ley . . . wanted me to modify or withdraw my remarks in fear that they might do damage to the cause of space flight." Ley's motivations are unclear. Did he believe that a more down-to-earth approach would damage the cause? It is possible, considering that so many of Ley's publicity activities promoted space exploration on a grand scale. Above all, his work for *Tom Corbett* indicated that he had few quarrels with ambitious visions of cosmic exploration. Nevertheless, there was a line that Ley would not cross.[24]

Perhaps Ley grew to dislike the tactics of von Braun and Ryan. Ley may slowly have disassociated himself from *Collier's*. Or there may have been a falling out. Daughter Sandra Ley recalls that Ryan was one of two names not to be spoken in the presence of her father. The other name was Walt Disney, for reasons unknown. Arguably, there was a deep tension between von Braun's efforts to promote grand visions of convoy trips to Mars and other disingenuous efforts to excite the public. Von Braun and Ryan may have crossed the line. Ley may have viewed von Braun as a salesman who, like Nebel, generated unrealistic expectations. Despite these possible tensions, Ley seemed happy to witness the ascent of von Braun as America's prophet and popularizer. It freed him to write about the past, instead of the future.[25]

Lands Beyond (1952)

In 1952 Ley published a book with co-writer L. Sprague de Camp, a friend and accomplished colleague. The book became a bestseller after it won the International Fantasy Award for nonfiction. In readable and entertaining prose, Lands Beyond explored the histories of myths, legends, and folklore. The authors unleashed their curiosity about mythical or mythological "lands beyond," while they reveled in scientific debunking. The book combined a reverence for the mysteries of nature with a heavy dose of skepticism.[26]

The authors began by discussing the "three colossal figures" of human history: "the warrior, the wizard, and the wanderer." While the warrior protected people from real foes, the wizard combated "supernatural dangers." Both the warrior and wizard led parochial lives. Meanwhile, the wanderer explored unknown places. He returned with "news of far and fantastic places." Each group had a vested interest in exaggeration. There were limitations placed upon the stories of the warrior and wizard: witnesses might dispute their most far-fetched tales. However, the wanderer "has a virtual carte blanche." The authors argued: "No wonder 'traveler's tale' has come to mean an elaborate lie or fantastic exaggeration!" Despite this knack for deception, the traveler was still an admirable figure. Travelers possessed "a burning curiosity as to what lay beyond the horizon." They sought out new frontiers.[27]

The wanderer "faced perils unknown even to him: devouring monsters, fierce people not altogether human, and the wrath of strange gods." Having proven himself as a fearless explorer, "he did not understate the dangers he had undergone." Instead, "by making the most of them . . . he could expand his ego, justify the high prices he wanted for his trade goods, and discourage possible competitors from horning in on his territory." Nevertheless, these tales "were seldom pure fabrications." Ley and de Camp untangled the real from the fake. They also celebrated the triumph of modern science, after scientific expeditions finally dispelled myths and legends. They demonstrated how "the worlds of geographical legend, after having been whittled down and shunted all over the map, have finally been pushed off the globe altogether." Modern science triumphed over myths, superstitions, and legends.[28]

Additionally, de Camp and Ley attempted a "restoration" of ancient and medieval worldviews. They appreciated how cultural beliefs and folklore shaped an imperfect understanding of the world. The authors put those worldviews in a cultural and historical context. They even compared themselves to paleontologists: "Still, as we can restore a dinosaur from its bones, so can we re-create these imaginary worlds from their traces in literature, folklore, and figures of speech." Not only was it possible "to write the story of exploration from the point of view of who discovered what and when," but it was also possible "to write the same story from the point of view of what people wanted or hoped to find." The history of myths and legends thus represented a history of human aspirations and fears. Although the modern world had been purged of demons and monsters, readers could still appreciate how premodern beliefs reflected human creativity, curiosity, and humility amid the mysteries of nature.[29]

Different chapters can be attributed to each author's specialized interests. For example, Ley's perspectives can be read in "The Land of Longing," his most forceful and entertaining debunking of Atlantis. Playfully, he discredited "careless . . . investigators, negligent in their logic, and given to believing whatever pleased them." These investigators clung to certain hoaxes "of the baldest kind . . . even after [they were] exposed." Much of the history of theories surrounding Atlantis included "logical slips so conspicuous that only hopeless credophiles could swallow it." The search for Atlantis had become a practice of cranks and "occultists who . . . pushed their way into the lost-continent domain." Ley wrote: "It seems, and actually is, impossible to make any sense out of such an enormous accumulation of supposition, commentary, cross- and counter-commentary and piled-on private beliefs and prejudices."[30]

In order to solve the mystery a rational person had to return to the only reliable primary source, which was Plato's *Timaeus*. Ley argued: "When we turn back to Plato himself we find that most of the perplexities just disappear." Atlantis was simply a literary device used by a philosopher who was "(to our mind) careless with the use of facts." The original story could still be appreciated. It revealed Greek literary devices, the political theories of an ancient philosopher, the circulation of Babylonian beliefs in catastrophe, and other historical gems. Ley concluded: "But the continent of Atlantis would never have appeared on any map of the real world, no matter when drawn."[31]

The book debunked many other legends. From the locations in the works of Homer to other "masterpieces of literary larceny," the authors exposed the lapses of logic, the unscientific reasoning, and downright charlatanism. De Camp and Ley even take on sacred legends of religious significance, such as the Christian kingdom of "Prester John" and the Jewish kingdom of the Ten Lost Tribes. These "lands beyond" only existed in "the cult mind." Each of these legends could be reduced to a simple historical fact or text. Often the easiest explanation was most valid. Nevertheless, these tales could still be appreciated for what they revealed about human longings for alternative worlds. The book's epilogue even laments the fact that none exist: "It's a little sad that they do not exist in fact." A mental tour of imaginative and romantic landscapes remained pleasant and educational.[32]

Reviewers praised the book. For example, columnist Charles Poore wrote, "Mr. Ley and Mr. De Camp rush up and down time's aeons with astonishing assurance." Poore further praised the authors' ability to destroy "all the beguiling moonshine." Nevertheless, a sense of loss could be felt, as the lands beyond "vanished, with so many other good stories." Poore then looked to the future by reviewing Arthur C. Clarke's *The Exploration of Space*. He argued: "It is a challenging link between the lost Atlantis and the unvisited geography of the stars." Another reviewer noted the theatricality of such a mental exercise: "Well, step right up, ladies and gentlemen, and meet the wonders of 'Lands Beyond.'" The journey was wondrous. The showmen were talented.[33]

Galaxy's Gig

In 1952 Ley signed an exclusive contract as science editor for *Galaxy*, a new science fiction pulp. Whereas earlier magazines like *Astounding* forced the science writers to compete for space, *Galaxy* wanted a regular expert for nonfiction articles. This role also fit well with the editor's desire to make the magazine far more respectable than an earlier generation of pulps. Not only would *Galaxy* stick to the scientifically plausible, but it would also focus on social issues, rather than technology worship. While other pulps endorsed pseudoscientific concepts, such as L. Ron Hubbard's "Dianetics," *Galaxy* would try to avoid such theories. It also paid writers more generously, which attracted quality stories. *Galaxy* published some of the most notable works by legendary science fiction writers of the "Golden Age."

Galaxy paid Ley $100 a month to write a "For Your Information" article on scientific subjects. The terms of this agreement required that Ley sever all ties to competing publications. In exchange, he would earn a regular income for a few days of work per month. This arrangement was a small part of a more lucrative five-year contract. Ley agreed to supervise *Galaxy*'s radio and television programs. Additionally, he would host a weekly radio talk at WJZ in New York City. For fifteen minutes on Sunday, Ley presented the program "Looking into Space." *Galaxy* paid Ley $75 for every weekly broadcast. He could also earn 50% of any net profits from the venture. In return Ley agreed not to appear on "any other radio or television program" without the prior consent of *Galaxy*. By this arrangement Ley could earn over $400 a month by writing a short article, spending an hour per month on the air, and serving as a general science editor. It is unclear whether his duties included vetting the content of stories. Thus the arrangement with *Galaxy* began as a very positive experience for Ley. Yet one day after signing the contract, publisher Robert Guinn tried to amend the terms to alter Ley's supervision of any programs "where we will not require your services." Ley may not have agreed to this alteration. He kept an unsigned copy in his personal files.[34]

Ley enjoyed his early association with the publication, in spite of an announcement for readers to "Send your science questions to Willy

Ley c/o Galaxy. He'll answer them all by mail or in this department [column]." For the rest of his life, Ley's mailbox at home and at *Galaxy* would be full of questions. Each month, he chose the ones he wanted to answer in a short feature called "Any Questions?" A typical column included a question about space stations, the natural causes of glacial melt, or the extinction of dinosaurs. Readers asked Ley to explain scientific theories. If he did not answer the question directly, he made readers do the math or measurements themselves. This type of direct reader interaction would continue intermittently throughout Ley's tenure. As the scientific expert in the pulp republic of letters, Ley promoted a participatory culture that had become a staple of the genre.[35]

Aside from this column, Ley wrote longer articles. His earliest contributions included pieces on astronomical subjects such as meteors or the question of "When Will Worlds Collide?" Ley poked fun at the publicity surrounding George Pal's new film, *When Worlds Collide*, and its loose connection to Immanuel Velikovsky's earlier theories. In other articles he returned to the subject of spaceflight. For example, one of Ley's top-billed articles predicted "Space Travel by 1960." The cover showed an enlarged V-2 inspired rocket, ready to launch to the moon. Ley commented on the progress of the last three years: "This issue's cover is something of 'instant recognition' to science fiction readers . . . [who] would have recognized such a picture even twenty years ago. Now, however, the same picture might be on the cover of any magazine." Ley added: "That is vast progress."[36]

In the next issue Ley moved away from space travel into the realm of general predictions. *Galaxy* used the cover to illustrate "The World of October 2052." It included a fashionable dinner party with robots and aliens. Despite the fantastic cover, some of Ley's predictions were fairly conservative. He did not believe "that the day of printed word has almost reached its end." Books would continue to be physical things in 2052. Ley also doubted the reliability of the "electronic device" that might someday replace newspapers. He viewed large cities and public transportation as the norm of the future, in spite of the threat of atomic warfare.[37]

He wrote similar articles for *Galaxy* for the rest of his life. These articles reached an expanding audience of dreamers. As *Galaxy* offered speculative fiction about the future, Ley steered the reader's imagination "in the right direction." He embraced his role as an expert on all things scientific. Ley appreciated how *Galaxy* promoted his role and his expertise. The editor announced: "Science had become so complex and confusing, even to scientists, that there must be some question that you'd like Willy Ley to explain clearly, authoritatively, and in everyday English." He continued, "As you can see for yourself, he's an expert in clarification. . . . He is not a scientific snob." The introduction added, "Now . . . what was it you wanted to know?"[38]

Ley taught readers about wondrous creatures, astonishing legends, and amazing historical facts. He also rewarded readers for asking questions. He encouraged their participation. He viewed science as open to every inquisitive amateur and hobbyist. In Ley's perspective, children and young adults held the keys to the future.

Disney in Space

When *Collier's* space-themed series ended in spring 1954, a greater opportunity arose. While some Hollywood brokers resisted the medium of television, Walt Disney began to embrace the genre, in order to promote and finance the construction of Disneyland in Anaheim, California. He organized programs according to the park's four themes: "Adventureland," "Frontierland," "Fantasyland," and "Tomorrowland." As Howard E. McCurdy notes, "Of the four themes, Tomorrowland was the least developed." Disney approached senior animator Ward Kimball and his assistants, saying, "You guys are modern thinkers around here . . . can you think of anything we can do on Tomorrowland?" Kimball had been reading *Collier's* magazine. He recalled, "It was really fascinating for me to realize that there were these reputable scientists who actually believed that we were going out in space." A space-themed program could mix imagination and science, in a way that entertained and educated audiences. Walt Disney expressed caution during a conference: "We should be careful and keep our serious stuff separate. We have to watch it so the material doesn't get corny."[39]

Allegedly, Disney handed a blank piece of paper to Kimball and said, "Write your own ticket." Suddenly, Kimball realized, "The key to his whole plan was the need to bring in prominent scientific advisors." Kimball aimed his sights on "qualified experts." Ley recalled: "The telephone operator told me that Burbank, California, was on the line . . . 'Walt Disney calling.'" Ley had to postpone the trip to California in order to attend a *Herald Tribune* cocktail party in honor of *Lands Beyond*. Then he flew to Chicago and boarded an overnight flight for California. In a later issue of *Galaxy*, Ley recalled the scene: "When I sat in the beautiful air-conditioned studios . . . I mentally weighed the problems involved." Whereas an author could assume some degree of education among readers who spent money on a book, a television writer had different considerations. Ley mused, "Obviously everything had to be explained right from scratch." He added: "On the other hand, the most instructive device invented so far was at our disposal: the animated cartoon. We would not have to *explain* with words, as I do in lectures; we could *show* how things work." After Ward Kimball explained Disney's storyboard process, Ley contributed to the initial idea of a program that explained the basics, traced the history of space travel in literature, and introduced viewers to scientific experts. The show would climax with a rocket flight to Mars.[40]

Two weeks later Ley returned to sign a contract and begin work. Disneyland historian David Smith described the scene as Ley had many meetings with different teams: "The men were fascinated with Willy Ley. Despite the odorous cigars that he chain-smoked, they gathered around him and hoped that some of his knowledge would rub off on them." An assistant recalled: "Willy was a real encyclopedia. . . . If you asked him a question, he'd pause for a second, then he'd say in his music hall German accent, 'Vell, as a matter of fact,' and then he'd take off with an encyclopedic description of whatever it was you were asking him. He was a very amusing fellow; we all got a big kick out of him." After several days of meetings, the producers agreed that there was simply too much material for a single show. They decided to make a two-part "cliff hanger," with the first part culminating in a trip to the moon, while the second part explored the space station and Mars. Walt Disney did not oversee many elements of the production. He was too busy with designs for Disneyland. However, on May 14, he made his intentions clear during a meeting, in which he said: "We are known for fantasy, but with these same tools that we use here we apply it to the facts and give a presentation. I think that's very important for this series—a science factual presentation." Disney also commented as to how the combination of experts and dreamers was "the key for the whole series." It would be exciting for the audience to see men "dealing with fantasy and men dealing with fact come together, meeting and combining their resources."[41]

At this point Ley recruited Wernher von Braun, whose attractiveness and expertise would appeal to an audience. Von Braun was reluctant to participate. He had competing arrangements with a Beverly Hills producer, and initially he turned the offer down. Historian Neufeld explained: "Disney did not give up so easily, largely because Willy Ley was its first hire and he kept von Braun's name on the agenda as the ultimate salesman and idea man for spaceflight." Eventually von Braun severed his ties to the competing producer, and the "Disney deal was on." He arrived in Burbank on July 10, 1954, nearly three months after Ley had been contracted. Ley positioned von Braun as America's rocket expert, while Ley took on the role of an educator.[42]

After von Braun and physicist Heinz Haber became more involved, many of the details changed. In retrospect, Ley recalled: "I don't know just what had been expected of the experts before they arrived; what we did do was to turn offices and sketch rooms into classrooms and apparently everybody was very pleased." Due to a wealth of material, the producers expanded the program to three parts. For the first episode, an entertaining and comedic cartoon would depict the history of the idea, while a middle portion would present the sober facts, as told by experts. This portion was followed by another humorous, yet educational cartoon about weightlessness. The show culminated in an animated trip around the moon.

Throughout the months of June and July Ley made additional trips to California. His consulting role may have expanded to Disneyland itself, as engineers designed and constructed Tomorrowland, which would include a virtual rocket trip to the moon and a space station where families could gaze upon the earth. Both Ley and von Braun were in high demand, as designers meshed visions of cosmic travel with sites of "Futurama."[43]

Engineers' Dreams (1954)

By 1954 Ley had already done much to advocate for large-scale investments in space travel. He had popularized scientific concepts, technological imagery, and narrative tropes surrounding the imminent conquest of space. These years marked the peak of Ley's celebrity status as America's authority on rockets and space travel. According to fellow science writer Isaac Asimov, Ley, "more than anyone else, prepared the climate within the United States for the space effort." Thus it may seem odd that Ley devoted several months to a book in which he wrote, "I am well aware of the fact that I haven't said a word about space travel." He claimed that he avoided the topic for the simple reason of writing so much about it elsewhere. However, the book reveals Ley's support for massive investments in large-scale technological feats.[44]

He titled the book *Engineers' Dreams*, an exploration of "great projects that could come true." Key examples included a tunnel linking Great Britain and France, the creation of massive lakes in Africa, and proposed efforts to drain the Mediterranean Sea partially. By tracing the histories of these projects, the book celebrated human ingenuity in the face of natural and political obstacles. The theme of the book was simple: these grand and expensive engineering projects were perfectly sane and reasonable. Yet they required international cooperation. Politicians must defer to engineers. Ley argued, "Engineers' Dreams are things that can be done—as far as the engineer is concerned. They are also things that cannot be done—for reasons that have nothing to do with engineering." Often these projects are impossible because the "sums involved may be so huge that only government could pay them." Convincing commercial interests to invest became an uphill struggle. However, the main culprits were "political difficulties." When an engineer's blueprint crossed the political boundaries of nation states, the project either died or became buried in "a mountain of paperwork."

The political troubles brought "too many highly uncertain factors." Consequently Ley could not predict the future of these engineering visions.[45]

In the past many of these projects had been considered fantastic. By the twentieth century, Ley argued, almost anything was possible. In fact, "the word fantastic, when applied to engineering, merely means 'it has not yet been done.'" Ley displayed an enormous amount of optimism surrounding future accomplishments. For example, it was "possible, in principle, to tap the heat of the earth's interior," thereby unleashing "an enormous untapped reservoir of power in the earth." Ley added, "The comforting fact is that the energy is there. If we need it, we'll find ways and means of going after it." Likewise, engineers knew that "a literally inexhaustible source of energy . . . comes every day, year after year, from the sun. All they had to do was to find a way to trap and harness it." Solar energy could solve the world's problems.[46]

Other projects were equally plausible and ambitious. If the nations of Africa united, it would be possible to flood the Sahara Desert by constructing a massive canal to the Atlantic Ocean. "In the end," Ley argued, "instead of a hostile desert you would have a large and navigable body of water." Regarding the native Africans in those territories, Ley declared:

> That the drowned area is an especially unhealthy place is generally conceded, but it is the home of a large number of Africans, about two million, who would have to be moved. Since the property of these Africans is mainly portable property, since the change would come about rather slowly, and, most important, since the move would certainly better their living conditions, it is unlikely that they would object.[47]

Ley also expressed confidence in the benefits of damming the Strait of Gibraltar and partially draining the Mediterranean: "The final result would be 220,000 square miles of new land and hydroelectric power plants of virtually unlimited capacity." Standing in the way of progress was "the political situation of today." Nevertheless, it was entirely possible if "a united Europe is a reality."[48]

It is easy to view these ideas as examples of mid-twentieth-century modernization theories. They promoted the violent redesign of nature, with little anticipation of unforeseen consequences. As a later reviewer recalled: "Ley's recounting of a cost-benefit analysis betrays the insensitivity of his times to issues that today evoke knee-jerk reactions." Overall the book illustrated Ley's modernism. He viewed the earth in terms of grand redesigns and the conquest of nature. Historians should recognize the more positive hopes within these grand visions. In his mind, these projects would benefit humanity. Ley expressed confidence that modern technologies could reshape the world. These projects would be acts of creative destruction that produced a world of peace and prosperity for all. New technologies could eliminate warfare, nationalism, poverty, and disease. Engineers could save the world.[49]

Still More Adventures of a Romantic Naturalist

In 1955 Ley published *Salamanders and Other Wonders: Still More Adventures of a Romantic Naturalist*. It followed in the footsteps of *Lungfish* and *Dragons* by investigating legends, myths, and case studies of "living fossils." With the vigilance of a detective, Ley "set out to find, if possible, the truth behind a few wild stories which are vaguely known to every naturalist (and are most decidedly known to be ridiculous)." "Let's find out what can be found," he stated. The book is still remarkable for its detailed investigation of sources.[50]

Ley spends much time debunking ridiculous ideas. Yet, at other times, Ley shows a remarkable curiosity and tolerance for the zoological mysteries of nature. For example, he begins the chapter on the "abominable snowman" by stating, "There is something unknown, or, at the very least, something unexplained." Ley then traces the history of eyewitness accounts. Often he lets the reader judge the credibility of a statement. He concludes with an optimistic statement about the yeti: "No matter whether Asia or Africa . . . the facts are that very primitive humans, sub-humans, and near-human creatures lived on both continents. And if a near-human and very primitive race shared the fate of the panda and the langur, its survivors would fit the description of the yeti perfectly."[51]

Ley could be associated with other "crypto-zoologists." He firmly shared the belief that there must be a basis for eyewitness accounts, which had their place in the history of science and exploration. In fact, the very essence of naturalism involved a quest to document and explain mysteries and legends. Ley made a direct analogy between these zoological quests and past discoveries: "Marvelous were the things that early travelers returning from the American tropics had to tell to their relatives and neighbors . . . doubts were in time crushed by evidence." The history of science included case studies, in which the explorer confronted baffling stories or evidence of strange creatures. Many explorers discovered wondrous creatures that should be extinct. "In short," Ley explained, "it happened so often that a tale, though exceedingly strange on the face of it, was proved to be true." Time after time, scientists confronted evidence of a creature that managed "to survive through centuries when nobody knew that it even existed." The naturalist must possess an open mind, along with a restrained humility in the face of the complexity of nature. To make a definitive claim about the unknown was preposterous.[52]

To date, *Salamanders* was his most widely read and critically praised book. It reached broad audiences, from scholars to teenagers. The book's broad appeal confused a few scientists as well as librarians, who had to determine how to classify the book. Was this scientific or historical, adult or juvenile, scholarly or popular? It was all of those things and more. In fact, the book's style led geneticist-turned-historian Conway Zirkle to review the work for *Isis*, a leading journal in the history of science. For Zirkle, two things dawned on him while reading *Salamanders*. Foremost, he was being "educated painlessly." Zirkle wrote, "The reader is apt to be startled by realizing that the information he has been absorbing is important." Zirkle then praised Ley as a pioneer in the field of the history of science: "Willy Ley is history-conscious and he presents each of his subjects in its historical context." Zirkle then made a passionate appeal to fellow historians of science:

All of Willy Ley's books on romantic natural history have ostensibly been written to entertain. On the surface they appear to be light literature—books to be read and forgotten. Actually, they are informative, scholarly and sound, and they emphasize aspects of the history of biology which most historians simply miss. Willy Ley shows beyond a reasonable doubt that scholarship need be neither solemn nor dreary, and that good scholarship, like good music or good literature, is basically enjoyable.[53]

Zirkle concluded by stating, "If historians of science wish to secure recruits for their discipline, a most effective means would be for them to present Willy Ley's books to some of the brighter high school students."[54]

Several scientists adored the book. For example, zoologist Fred R. Cagle admired the "daring of Willy Ley." He continued, "It is striking that Willy Ley, a free lance writer, managed to present more, substantial zoological knowledge than many zoologists have done in their popular writings." Cagle recommended that other zoologists emulate Ley. Ornithologist Burt L. Monroe called the book "serious scholarship and research." Other reviewers praised Ley's style and research methods. A writer for the *Nation* argued that Ley "has a bloodhound's persistence. . . . Give him a vulgar error or a lost animal to trace down the centuries and he is happy." Writer Thomas Gardner further praised the book: "No writer has excelled Ley in describing both real and mythical things in nature, using a scholarly background in such a manner that the reader learns while enjoying the broad knowledge of the author." Gardner added, "Mystery, legend, and fact are blended together in a romantic form." Critics appreciated the common themes that united Ley's works on rockets and the mysteries of nature. According to a reviewer, "Willy Ley is equally at home among prehistoric animals or space ships."[55]

Salamanders was also described as a travelogue. It illustrated how Ley's books were literary "voyages of discovery," according to August Derleth. His landscapes were alien. The journey was bold. Readers co-experienced the moments of wonder, awe, and sheer amusement. The book displayed the "enduring wonders of the world around us." Ley's enthusiasm was infectious. This book was "science writing at its very best." Ley turned a dull subject into an "unforgettable pleasure."

Yet another commentator noted how Ley's romantic naturalism provided "after-dinner conversation of the next six months." People were
talking.[56]

Tomorrowland and Beyond

Ley and von Braun's roles as consultants to Disneyland's Tomorrowland began to materialize, as the construction of the theme park
neared completion. According to a reporter for the *Chicago Daily Tribune*, one of the most notable exhibits and experiences surrounded a
mock rocket ship of the future, along with a space theater that would
"take visitors to the moon and back." Visitors would marvel at the "76
foot rocket ship pylon, painted gleaming white with TWA's characteristic markings," before entering a large theater to experience the
everyday trip to the moon that would happen "about 30 years from
now." After passengers were greeted by TWA attendants and pilots,
they would take their seats for the journey to the moon: "Two scanning
screens—one in the base and the other in the ceiling—will present a
picture of what a passenger would see on a flight to the moon. Sound
effects, air and temperature changes, and mechanical movements of
the seats will confirm the illusion." Voyagers watched with wonder
and awe as the earth retreated below, while the moon expanded above.
The seats deflated to give the sensation of velocity. It was more than
a simple amusement ride, according to commentators. Instead, Disney combined "animated cartoon style and the newest in three dimen-
sional realism." The ride was based on "the best scientific advice and
knowledge available."[57]

After making the journey from the earth to the moon, families could
then tour "Space Station X-1." From a perched balcony, they looked
down upon the earth to gain a "Satellite View of America," according
to Disney promotions. At a height of 90 miles and a speed of 60,000
m.p.h., "the round room slowly revolved around a beautiful, detailed
landscape."[58] The audience would experience Futurama at the most
extreme heights. Although it was a less popular and educational exhibit, it must have inspired wonder, awe, and enchantment. The future

of Americans in space was guaranteed. As Disneyland put the final touches on this exhibit, Ley happily took his place behind the curtain to speak through the character of Tom Corbett, Space Cadet. The actor channeled his voice, just as science fiction, media, and rocket rides had done.

Facing the Nation

In late July 1955 Eisenhower announced the intention of launching a man-made satellite within the next two years. Journalists and other commentators immediately turned to Willy Ley to explain the announcement. Ley argued that the decision "opens the age of space travel." As the "first step in space," the American satellite would pave the way for better satellites. Ley stated, "In principle, the problem is not very difficult." Ley also predicted the future: "The third or fourth [satellite] may well carry a television camera to show us what the planet Earth looks like when seen from space. . . . By that time, a manned rocket ship will go into an orbit around the earth and after that engineers will begin to plan manned space stations." He concluded, "The artificial satellite is going to be a major accomplishment, but its main importance will be that it will be followed by others. . . . And after that, in time, there will be a manned artificial satellite that will travel through space." Ley made no mention of a space race with the Soviet Union.[59]

On August 14, 1955, Ley sat down in front of cameras for a long interview with CBS's *Face the Nation*. The panel included William Hines of the *Washington Star*, Carleston Kent of the *Chicago Sun-Times*, and Erik Bergaust of *Aero Digest*. Host Ted Koop introduced Ley as a "leading lay rocket expert." Hines immediately launched into a question surrounding conflicting reports about a rocket already launched from White Sands. On the one hand, authorities had announced the intention to launch a satellite. On the other hand, reports circulated that this had already been achieved in secret, which the Pentagon denied. Hines asked, "What can you tell us about it?"[60]

Yet again, Ley found himself in the limelight. Although he was well removed from the centers of research and experimentation, as well as the avenues of military hierarchy, he had to discuss his perceptions of events. He was, after all, an expert on rockets, missiles, and space travel. "My personal reaction is," he stated, "I don't believe it, for the simple reason that getting a satellite into space is a possible but fairly complicated activity, and I do not think that this could happen by accident." Hines asked, "Well, do you think it could happen by design, to an altitude of 250 miles, which they are talking about now?" "Oh, certainly," Ley casually responded. He added, "We were in the theoretical stages about ten years ago. Ten years ago it became perfectly clear that it was not only feasible, from an engineering point of view, of sending a small satellite into an orbit around the earth; it was also clear that this would be a very useful experiment[,] that we would learn a great deal from it." Ley also stated, "by now . . . we should have a satellite carrier."[61]

Kent then asked the big question: "Mr. Ley, what is this all going to lead to? What good will be accomplished in the whole thing?" Ley answered with a brief discussion of satellite data as necessary for the future conquest of space. Kent responded, "Then you're saying, in effect, that the major finding, out of this whole project will be to enable us to build space ships in which people can travel?" Ley confirmed, "This is precisely what I said, although I did not use the words, but this is only one of the results. In the meantime, we learn a lot of other things." After side-stepping issues surrounding the military implications, Ley then responded to a pressing question: "Well Mr. Ley do you feel that the Russians are really on par with us in rocketry?" Ley answered, "In general, my feeling is, here, that the Russians can do it as well as we can, but that we can do it earlier, or faster or better, or all three." Ley then added his skepticism regarding Russian announcements: "Well, there is one thing with announcements coming out of Russia: You never know whether they are announcements, propaganda gestures, tests of public opinion, or whatever." At the conclusion of this interview, Kent asked, "Do you still think it would be valuable to go to the Moon?" "At least once," Ley answered. "It would be of scientific and possibly of practical value to go to the Moon at least once."[62]

Following this interview Ley embarked upon another lecture tour. Then he spent much time revising and correcting his *Rockets*. He told Heinlein that supplies of existing editions were running low, with an average sales rate of one hundred per month. He then joked, "Wonder how many revisions I'll live through." Ley would soon turn fifty years old. His celebrity status peaked.[63]

The *Sputnik* Challenge

In a 1957 interview Ley announced, "I can predict the future when it comes to machinery." He described the future of space rockets, satellites, and space stations. He had been making such predictions for most his career. In the mid-1950s Americans listened to his voice or read his words in newspapers. Prior to the launch of a Soviet satellite in late 1957, Ley reassured people about future American accomplishments in space. He thrilled both adults and children. Much of the audience eagerly anticipated the grand opening of a new frontier. Ley continued to celebrate the scientist-explorer. Meanwhile, von Braun shared the stage as a potential hero and rocket expert. Ley relentlessly promoted von Braun as the man who could guarantee a future of missiles, satellites, and space travel.[1]

After the launch of *Sputnik*, Ley further reassured people. Yet his tone became far more anxious and urgent between October 1957 and January 30, 1958, when von Braun launched the first American satellite, *Explorer I*. Many of Ley's predictions had to be modified. He stressed the urgency of the moment. As the public perceived a missile gap, Ley lamented a cultural lag with the Soviet Union. He also debunked false claims about Soviet-German rocket engineers. Just as Ley had a duty as a science writer during the Second World War, he now had a duty to educate the public, when he witnessed public panic and controversy. By documenting his perceptions, activities, and tactics, we can explore the "shock" of *Sputnik*, as the press portrayed it.[2]

The Exploration of Mars (1956)

Both Ley and von Braun had been voicing predictions for several years. Recently their co-authored *The Exploration of Mars* (1956) depicted an ambitious first voyage to Mars with two ships and twelve men. Granted, the book presented a scaled-down version of von Braun's earlier designs for a massive twelve-ship convoy. Nevertheless, the book communicated a vast sense of optimism regarding human spaceflight. Some of the descriptions of the voyage read like passages from a *Tom Corbett* story: "Suddenly, the tense exchanges become downright nervous. We hear the engineer yelling something into the intercom, and the captain yelling back. One of the lights on the engineer's panel is flickering a red warning as he fumbles excitedly for the correction switch."[3]

The mysteries of Mars would soon be solved. In flowing prose Ley anticipated the moments of discovery, after outlining a long history of dreams and assumptions about the red planet. Far more than his earlier *Conquest of Space*, the book on Mars delved into the history of science and the bold explorers who dreamed of first contact with Martians. Over half of the book recounts the history of observations and theories, as Ley and von Braun take readers through journeys of discovery. They conclude: "We, the genus homo of earth, will set foot on Mars within a matter of decades." An expedition to Mars was a logical part of a "step-by-step development." Bonestell's illustrations complemented these predictions. From a photographic depiction of the earth's first satellite to the manned colonization of Mars, Bonestell's painting illustrated the future.[4]

Reviewers loved the book. For example, science writer John Pfeiffer stated: "Their latest imaginary adventure is spelled out to the finest possible detail on the basis of modern space technology." The *Wall Street Journal* noted, "Once again, Willy Ley and Wernher von Braun have hurtled the future into our laps . . . these indefatigable rocket experts have presented a down-to-earth 'how-to' master blueprint of

Earthman's exploration—'within a matter of decades'—of the planet Mars." This book did much to fuel public expectations for an immediate conquest of space. There was a logical step-by-step progression, when it came to Americans in space. The book showed little awareness of a space race with the Soviet Union.[5]

American Firsts

Ley also predicted the successful launch of American satellites during the International Geophysical Year. Allegedly, these satellites would lead to a manned rocket and a space station. Ley glorified a new era of human exploration. For example, a widely read issue of *Science Digest* communicated his predictions. Ley argued that the "really big headlines will come in a year or so, when the first Vanguard three-stage rocket roars into the sky . . . to carry the first American artificial satellite into an orbit around the earth." According to Ley, Project Vanguard would unleash a wave of scientific discoveries.[6]

Vanguard would be one of many shots "designed to ferret out another set of unknowns." After Vanguard, an atomic or solar powered satellite would use a television camera to peer down upon the earth. Within ten or twelve years a network of satellites would broadcast television. By then the problem of re-entry would be solved, which "will open the gate for two developments." Ley explained: "One would be the long-range rocket-propelled passenger liner . . . which could cross the Atlantic outside the atmosphere in not quite an hour." Second, three-stage rockets would ascend with components of a space station, to be constructed quickly in stable orbit. Ley predicted the future based on his perceptions of the past: "Just about two decades elapsed between the drawing of the preliminary plans for the first large rocket to the drawing of the plans for Vanguard. Between the first Vanguard shot and the assembly of the first manned space station another two decades may elapse. But the time in all probability will be shorter." Again, Ley seemed unaware of any degree of competition with the Soviet Union.[7]

He made similar predictions in the pages of *Galaxy*. For example, in "Between Us and Space Travel," Ley outlined three main obstacles to be overcome: the "re-entry problem," the physiological effects of "cosmic rays," and the toll of zero gravity on the human body. On the first issue of re-entry, Ley expressed optimism: "It is rather safe to say that the re-entry problem is not solved at this moment. But engineers feel sure that they can solve it, provided they have exact and reliable figures to work with." Ley added, "The artificial satellites will provide these figures." Additionally, the satellites would help solve the problem of cosmic rays by measuring their effects. On the difficulties, Ley simply argued, "The answer is that we don't know yet. . . . But we'll find out. And when that has been done, another barrier between us and space travel will have been removed."[8]

Ley made other predictions on radio and television broadcasts. He also gave many public lectures. Venues included Rotary clubs, hotels, museums, churches, and radio institutes. Most notably, he informed an audience of navy reservists that "all of the materials and mechanisms needed by engineers to build the Vanguard were known and in use by 1900, but there was no theory that said it could be done." He probably gave a similar, hyperbolic lecture at the annual meeting of the Amateur Astronomers Association. Ley voiced his confidence that Vanguard would launch the first satellite into space in late 1957. He presented the man-made satellite as an American accomplishment, destined to succeed.[9]

Rockets, Revised

In 1957 Ley revised *Rockets, Missiles, and Space Travel*. By this point the book was a bestseller. Exact sales figures are unknown, but based on a survey of correspondence, the public demand for Ley's book rose steadily between 1954 and 1957. Ley's 1957 edition of *Rockets* went through six printings by fall 1958. Additionally, reviewers and advertisements labeled it as "the" definitive book. It was at this moment, to quote historian Roger D. Launius, that the book became "one of the most significant textbooks in the mid-twentieth century on the possibilities of space travel."[10]

Ley's changes to the text are telling. He entirely removed the chapters "The Meteorological Rocket" and "Terminal in Space" to make room for a new chapter on satellites, titled "The Shot Around the World." The chapter began with a reassurance regarding predictions. Ley wrote, "I have repeatedly said that I know of no other science which has such a magnificent record of living up to its own predictions as rocket research has had." On the satellite, Ley answered the public's central question of "What holds it up?" Much of this discussion anticipates the success of Project Vanguard. Ley does not comment on possible Soviet achievements. The unmanned satellite is presented almost entirely as a new idea, first inaugurated in army circles, and then put forward more directly by Ley in a 1949 article for *Technology Review*.[11]

This presentation is a curious contradiction to the 1949 article, which presented the idea as "not really new." In the original article, Ley mentioned a 1946 report by Defense Secretary James Forrestal. Ley attributed the germ of the idea to German pioneers: "The German literature of the pre-Hitler period dealt with manned artificial satellites, which were intended to serve as refueling stations. . . . An artificial satellite was 'obviously' useless without at least one observer, and an observer could reach the satellite only by piloted rocket." By 1949 Ley's attitude had changed: "Because we now have instruments which not only register their readings but which can also report their findings by radio, a satellite rocket can be quite useful without a living observer."[12]

Ley's new addition to *Rockets* took the case further. The book also reflected upon the broader scene. Suddenly newspapers and magazines buzzed with talk of satellites and space travel. Ley spoke directly to future historians:

It would not surprise me too much if somebody in the future tried to make out a case that during the years 1953 and 1954 a number of "space-happy" scientists . . . carried out a conspiracy to talk their government out of tax money for their wild schemes. . . . To discourage a possible future compiler of such a story at the very outset I can tell him that it all more or less just happened. The dozen or so men who talked space travel had talked space travel all their adult lives, but the time was ripe and they had, quite literally, bigger and bigger opportunities for talking.[13]

Ley added, "One thing had simply led to another." Public excitement for American spaceflight was at its peak intensity. Adults and children clamored for information.

Sugar Jets in Space

In mid-1957 Ley entered into an agreement with Sugar Jets Cereals, which unleashed a massive marketing campaign. They offered Space Age rewards to children who located special boxes of Sugar Jets. This campaign also centered on spaceflight television commercials that aired during *The Mickey Mouse Club*. Ley (along with physicist Joseph Kaplan, teacher John Sternig, and artist Chesley Bonestell) advised the producers of these futuristic commercials. Actor John Fink described one of the ninety-second commercials as "one-third advertisement and two-thirds information about rockets and space travel." They depicted the immediate American conquest of space. Sugar Jets also made an "amazing double offer." If children sent a special box top, plus $1.00, Sugar Jets delivered an eighteen-inch-long telescope, which could be used to "see the flight of the first man-made satellite that may be circling the earth this very year!" Young stargazers also received a "Sugar Jets' Space Map." The second offer related to a new and exclusive book by Ley, titled *Man-made Satellites*. A Sugar Jets fan could order this "complete easy-to-read account of the soon-to-be-launched man made satellites!" An ad proclaimed it "Not fiction. The real story." The ad then announced: "Yes, this year start your world of the future—your world of adventure—*your world of outer space!*"[14]

Ley's *Man-Made Satellites* was only forty-four pages, with thirty-two illustrations. The first sentence of the book read, "This is the Special Events Division of your local station. We are now switching you to Patrick Air Force Base in Florida for the firing of the first artificial satellite. . . . Take it away Florida." A reporter on the scene informs the audience about Project Vanguard, which was "the shot around the world." The countdown begins at "X minus 20 minutes," as scientists and engineers signal their status as "ready." The final thirty seconds "seem to take longer than thirty minutes in the outside world." The text culminates in a dramatic lift-off: *"One . . . Zero . . . Fire!"* Spectators marvel, as the rocket ascends. Ley announced: "Earth's first artificial satellite is on its way."[15]

The remainder of the book explains the physics of the rocket's ascent and the satellite's orbit. Ley also predicted a permanent American satellite. This satellite would save lives by giving meteorologists advance warning of extreme weather. An illustration depicted a weather control center of the future. The book ends with a dramatic image of a manned spaceship. Ley wrote: "In less than a decade some men will look through the pressure-proof and radiation proof windows and see the wide curve of the earth below. They will look up and see that the sky is black. And they will know that they are in space."[16]

This book was enormously popular with young readers. A journalist recalled, "The demand has been tremendous." The success prompted General Mills to commission immediately a second book by Ley. Published as quickly as possible, *Space Pilots* described "the *people* who will fly the rocket ships." Ley added, "Some of you may be among those who will be trained for this task." After exciting the readers with a dramatic launch sequence, *Space Pilots* describes what it takes to make the grade at space academy. Unlike the twenty-fourth-century academy of *Tom Corbett*, this Space Academy, U.S.A., was just beyond the horizon. The selection process would be unforgiving. The cadets must be physically and mentally fit. They must be exceptional pilots as well as mechanics. Experts on the ground would do much of the work. Still, it would be crucial for space pilots to possess expertise in complex mathematics. They must be exceptional students: "When he is finished with his studies he will know a great deal of engineering; he will have mastered some branches of astronomy; he will know a great deal about higher mathematics; he will have studied some aspects of medicine. . . . All this in addition to having become a pilot." The book illustrates these points with a picture of a young and clean-cut Caucasian man, laboring through a stack of large books.[17]

Space Pilots then turns to individual topics, such as the period of ground testing that the pilots would endure. Ley predicted a wide use of simulators and stress tests. He also predicted a transition from the experimental X-49 to passenger liners that could travel from New York to London "in about one hour," if the trips could be made affordably.

Ley then ruminated on a space station, which would be manned with research scientists "investigating the laws of nature." He argued, "Some day passengers on transcontinental or trans-oceanic flights will see the space station overhead, and will wonder at that moment what the researchers are doing to change their lives." Ley further advanced these themes in two subsequent General Mills books.[18]

Night-Beat

In July 1957 Ley sat across from journalist Mike Wallace for an episode of *Night-Beat*, a television show that soon aired nationally as *The Mike Wallace Interview*. Wallace had already gained a reputation as a provocative journalist, particularly after a tense interview of the imperial grand wizard of the Ku Klux Klan. Wallace asked provocative questions. Often his guests squirmed in their chairs.[19]

"Now to another world," Wallace announced, "literally . . . to the second story of rockets, satellites, and flying saucers with science writer Willy Ley." Wallace added: "Born in Germany, Willy Ley took degrees at the universities of Berlin and Konigsberg." Apart from his birthplace, this information was false. Yet Ley did not correct the host. Wallace continued: "[He] planned to be a geologist but switched to rocket research and was one of the founders of the German Rocket Society in 1927." Again Ley did not correct the host, even after Wallace announced that Ley had three (instead of two) children. Almost immediately, Wallace asked his most pressing question: "Willy, you left Hitler's Germany in 1935. Other men stayed on. . . . You got out. You came here the hard way as a refugee. They stayed, worked through the war, and then came over to our side. How do you feel about their switch?"

Ley let out a sigh, adding, "Well, it is of course hard for me to tell how I feel about somebody else's switch." Ley characterized the rationalizations of Germans: "They simply feel themselves, now that the world is divided into an Eastern and Western hemisphere . . . they feel themselves as Westerners." Wallace pressed, "But what about the switch, Willy? They pulled one to some degree." Ley defended the group: "Well, it was partly pulled on them." Ley presented von Braun and team as unemployed scientists in 1945. Fearing for their lives, they approached the Allies. "They had no idea how the Allies would treat them," Ley added. Wallace interrupted, "Do you know them well today?" Ley responded: "I know them well, yes. I can say that." Wallace asked: "When you get together, what do you talk about? The Nazis? New times in America?" "Ehhh . . ." Ley answered, "We practically never talk about politics." Wallace then inquired: "How do these men reconcile having worked for Nazi Germany, building rockets against the Allies, and then turning around to serve their former enemies?" Ley cringed, before stating that the group did not think of the Western powers "so much as enemies, personally." Wallace interjected, "Why didn't they get out as you did?" Ley responded: "I was a young man. I didn't have any ties. I could manage." Other people had property or positions of importance. Ley added: "I never officially left. I disappeared."

Wallace pressed Ley on the patriotism and national loyalty of the German rocketeers. Ley stated: "Much of it was probably just routine. You see you have people with strong political feelings and you have others who don't." Von Braun and his team were apolitical engineers, according to Ley. Wallace then asked, "Suppose we were to fight a war with Russia and we lost it. How do we know they wouldn't pull the same switch?" In a frustrated tone, Ley answered: "How do we know anyone wouldn't pull the same switch? Personally, I doubt it. But who can predict the future, when it comes to people?" Ley only predicted the future when it came to machinery. For the remainder of the interview, he expressed confidence in the imminent American conquest of space.

American Missiles and the "Shock" of *Sputnik*

Despite his optimism about American firsts, Ley recalled that 1957 "was a year that offered no comfort to Americans." Although few details were publicized, Ley followed the failed launches of the Thor missile. He was greatly surprised by Thor 103, which "exploded on the pad at T-4 minutes!" Ley also followed reports about the Atlas 4-A, which exploded one minute after launch. Ley remembered, "The public was dismayed, to put it mildly." On the role of the press, Ley remembered: "newspaper reporters, editorial writers, radio commentators took the position that such a big rocket simply could not be made to work; they did not mention the fact that about half a century earlier some airplane crashes had also been taken to be proof that 'flying machine' could not work." By August the scene of American rocketry was deeply discouraging. According to Ley, "even those who knew all the events" had good reasons to be disappointed. Ley later described the scene: "No Thor missile had made a successful flight, a Jupiter missile had malfunctioned, and the only Atlas flown had exploded. . . . The Russians chose this month to announce that they had an ICBM [intercontinental ballistic missile] . . . it deepened the gloom. And no relief was in sight." Ley recalled a series of other failures throughout September.[20]

When the Soviet Union launched *Sputnik I* on October 4, 1957, Americans looked to the sky. Some parents may have borrowed their child's Sugar Jets telescope to look in vain for the object. Others tuned in their radios to hear hourly reports of the satellite's current location. If historians take the press accounts at face value, many Americans had questions and concerns. This success in spaceflight technology represented a triumph of the Soviet system in the realm of science. Journalists and scientific intellectuals expressed a sense of shock, which they communicated to the public.[21]

In order to understand the tone of this reaction, as well as Ley's subsequent activities, it is important to stress the themes of earlier chapters. Many scientific intellectuals, including Ley, had contrasted a scientific democracy with a totalitarian regime that would invariably stymie freedom of thought. According to this perspective, the Soviets lacked the engine that drove science and innovation forward. Many scientific thinkers would have agreed with J. Bronowski's popular 1954 book, *Science and Human Values*, in which he argued that the "spectrum of values" of a scientific society included dissent, freedom, and independence. Bronowski also claimed: "No one can be a scientist, even in private, if he does not have independence of observation and of thought."[22] In contrast to the totalitarian societies of the East, science flourished in the West due to freedom and democracy. Bronowski argued that science simply could not advance in a dogmatic society that restricted human freedom. He claimed, "The society of scientists must be a democracy." *Sputnik* called these widespread associations into debate. Totalitarianism should have stunted innovation and progress. Totalitarianism should have hindered the great leaps forward. *Sputnik* should not have happened.[23]

Soon many scientists rushed into the classrooms of America. As historian John Rudolf documents, their educational efforts involved a fascinating combination of hope and fear. Scientists and science teachers voiced an incredible amount of hope that the American public could be better trained, not only in the fields of science, but also more generally in the realm of scientific thinking. With education as a "central plank in their movement," they sought to reform the American citizen and (by extension) the American system. The public understanding of science was crucial, because it not only combated public misperceptions but also justified an improved social standing for the scientist. This standing in turn justified massive centralized funding. John Rudolph illustrated how many of the associations between science and democracy began in the 1930s. More recent scholarship has pushed the associations further into the past. Regardless of the exact origins, it seems clear that by the early 1950s many scientists, educators, and intellectuals generally contrasted science and democracy with pseudoscience and totalitarianism. Technology, in their minds, was the ultimate manifestation of applied science, thereby ensuring that the United States was destined to lead the world in modern marvels.[24]

Sputnik undermined many of the assumptions about totalitarianism and pseudoscience. *Sputnik* frightened Americans for other legitimate reasons. Scientific intellectuals expressed outrage. Citizens asked questions. Journalists and experts reported. The media response was not simply an unjust attack on Eisenhower by opportunists, nor was it simply a dishonest media frenzy. *Sputnik* challenged Americans to reevaluate the relationship between science and democracy. Soon the very voices that had once praised decentralized competition now called for centralized planning and oversight. Many of these individuals rushed to a microphone or newspaper to counter public anxieties, as congressional legislation led to the creation of NASA, expanded funding for the National Science Foundation, and other very expensive reforms, such as the National Defense Education Act.

A Media Frenzy

Many journalists turned to Ley for answers and explanations. As a journalist noted, "Willy Ley was one of the first scientists to whom the newspapers turned for explanations." Requests for interviews and articles poured into his mailbox, along with the questions of *Galaxy* readers. In a letter to Heinlein, Ley described his daily routine: "Yesterday I had the first non-business lunch since October 4th . . . and over the weekend I'll have to produce some 5000 words of copy plus about 20 letters . . . Sputnik zemli has brought me a lot of cash, I must say that for it." He began answering questions, as best as he could.[25]

Initially, he shared the public's shock and doubt. Nevertheless, as Lester del Rey recalled, Ley "shrugged and went cheerfully on television to calm and explain and offer hope." For example, he joined a small panel on the CBS television show *Eye on New York*. He stated that the United States could have launched a similar satellite a year ago. He added, "but it wouldn't have done much good," due to the lack of a global system of observation. Ley argued that "we (the United States) have been beaten only in the sense of propaganda value." According to the *Chicago Daily Tribune*, Ley was also asked if the Russian satellite was made possible "because the Russians had the right Germans." Ley responded: "This is almost certainly wrong unless the Germans had rocket experts they didn't know anything about." Ley spent much time debunking these rumors. He also appeared on ABC's special report, "The Red Satellite." He similarly argued that the United States could have launched a satellite a year before, yet such an accomplishment was pointless. The only value of the Soviet launch was propaganda. Following the launch of *Sputnik II*, Ley appeared on other television shows. In one program he attempted to dispel rumors about the Soviets' ability to bring an animal safely back to earth. He also suggested that the Russians might attempt to launch a moon rocket to inaugurate the fortieth anniversary of the Bolshevik Revolution. He still voiced confidence in American technology, yet his tone became more cautious and hesitant.[26]

For the next six months Ley campaigned to both educate the American public and debunk the propaganda. Just as the Second World War compelled him to write "war weapons" articles for *PM, Sputnik* compelled Ley to write for newspapers again. He enlisted in a fight. At precisely this moment, Ley joined the *Chicago Sun-Times* as a "staff writer." When the *Sun-Times* announced his appointment, it ran a full-page display ad that included a portrait of Ley and a satellite hovering above his head. The ad read, "On the Threshold of Space. . . . To keep you authoritatively informed as we enter the Space Age, the Sun-Times is proud to announce the appointment of Willy Ley." The ad described Ley as "a noted scientist in the field of rocketry." The details of this arrangement are unknown. Most likely the *Chicago Sun-Times* contracted Ley for exclusive articles. The paper syndicated copies to regional newspapers, like the *Los Angeles Times* and the *Houston Post*. According to a short biography, these articles were syndicated "to more than 100 newspapers." When Ley had something to say about rockets, hundreds of thousands of Americans read his words. Not only were these articles distributed in newsprint, but several also ran in the magazine *This Week*. Additionally, Ley's pieces were so popular that New American Library collected many together into a paperback book for mass consumption. With a price tag of thirty-five cents, *Satellites, Rockets, and Outer Space* gave Ley's articles a wider readership.[27]

Ley's first series for the *Sun-Times* was titled "Missiles, Moons, and Space Ships." Topics ranged from "What Will Invaders from Space Look Like?" to "Von Braun Was Rocket Pioneer!" On the one hand, these articles continued to voice optimistic predictions about the American conquest of space. His boldest prediction asserted that manned colonies on the moon and Mars were just beyond the horizon. Other predictions outlined the imminent series of steps that would lead from satellites to a manned space station. Additionally, Ley predicted: "If we visualize a well-planned and well-co-ordinated Operation Outer Space the first orbital flight of a manned ship looks as if it were six or seven years in the future." He claimed, "Space travel will follow as naturally as air travel followed man's first winged flights." Allegedly, Ley also forecasted the future for the attendees of an annual charity event. A journalist noted, "Interplanetary travel will become so commonplace, commuters may find it a bore, Willy Ley, rocket expert said yesterday."[28]

In spite of this optimism, Ley's tone became more cautious and urgent. For example, he also attended several youth forums, where he gave talks such as "What Is Man Seeking in Outer Space?" A specific event was televised. Ley argued that the United States did not lag behind the Soviet Union in terms of general science. Rockets were a different story. A journalist reported, "Mr. Ley . . . said the United States was behind by at least a year in artificial satellite work and in the development of missiles of intermediate (1,500-mile) range." There was a missile gap, according to Ley. On December 6, 1957, the attempt to launch Project Vanguard failed when the rocket exploded on the launch pad. Ley recalled the aftermath for the readers of *Galaxy*: "I was subjected to more than the customary number of radio, television, and newspaper interviews, not to mention countless private questions. They all dealt with Vanguard, which has just suffered the most publicized failure of any rocket." Ley added, "And since everyone had been whipped into expecting wonders of Vanguard, the disappointment was obviously severe." Ley overlooked his own contributions to the optimism surrounding Vanguard. He began to caution the American public to expect more failures.[29]

As accounts of the failure spread throughout the press, so too did a *National Review* article by Ley: "Some Implications of the Sputniki." Ley began by commenting on so many contradictory accounts. "Ever since Sputnik No. 1 took up its orbit around the earth," he argued, "thorough newspaper readers must have been thoroughly bewildered by the headlines they could read in succession." He joked, "One read: 'Sputnik's Meaning: 'Catch Up Or Die' Says Rocket Engineer.' He was immediately fired from his job. Another one was 'Just a Silly Bauble Says Presidential Advisor.' He was not fired." Ley then tried to set the record straight by debunking false rumors. The Russian satellite itself had no direct military value. However, Ley argued, "The military significance of the Sputniki is in the rockets that launched them." He presented the facts to the best of his knowledge. The Russians had a 1,500-mile rocket. "We do not," Ley emphasized. The Russians were testing 4,000-mile-range missiles. Ley speculates, "We may have; the public has not been told." Ley concluded, "Our job is very simply to catch up with the Russians." He made a similar statement in an article titled "Soviets Seen Far Ahead in IRBMs [intermediate-range ballistic missiles]."[30]

Ley never missed an opportunity to promote von Braun as the man who could lead Americans into space. Von Braun's fame had been increasing since the *Collier's* articles. After von Braun successfully launched the first American satellite in late January 1958, he became a national hero and scientific celebrity. Historian Neufeld explained: "Within four months . . . he was a bona fide American hero, the Western world's most prominent gladiator in a celestial contest with the Soviets." Von Braun took center stage as a promoter, salesman, and publicist. His image soon graced the cover of *Time*. Other magazines clamored for autobiographical articles. This demand eventually inspired a movie, *I Aim for the Stars* (1960). As von Braun's fame exploded in 1958, Ley glorified his accomplishments in newspapers and magazines. Soon they would be co-attending events.[31]

Meanwhile, Ley continued to celebrate the scientific and technological possibilities. He predicted television, meteorological, and communications satellites within the next two years. Satellites would usher in a "new communications era," when "messages which came from the United States will be released on demand" around the world. He also outlined two methods that could be used to launch a satellite to the moon in 1958. He wrote: "One doesn't have to be a prophet to predict that there will be a shot to the moon this year. Nor does one have to be a cynic to say that there will probably be two—one American and one Russian."[32]

Ley claimed that the "first moon trip" could happen as early as 1965. Whereas earlier predictions saw the establishment of a space station as a necessary precursor to a moon voyage, Ley argued, "The trip around the moon . . . doesn't have to wait for the establishment of a space station." In some articles he argued that the moon should be "Target No. 1." Yet other articles continued to promote von Braun's space station. Reporters eagerly informed the public of his predictions of a "space station, a red, white, and blue one." Ley also promoted the future "passenger liner." In an article for the *Rotarian*, he celebrated the "point to point" rocket flights that will be "manned and peaceful." Ley made other predictions while informing readers about American progress in space.[33]

Simultaneously, Ley expressed anxiety and fear, which increased with each Soviet success. Ley began to think in terms of cultural lag. It started with a comparison of the public reception of Goddard and Oberth. Goddard's original treatise "was received with a small amount of ridicule," and it generated no noticeable scientific debate, Ley argued. Conversely, Oberth's work "was received with complete seriousness. . . . And it started things going on an international front." This comparison led Ley to imply that the negative public reaction stunted the field of American rocketry. He took this argument further: "We could be far ahead of the Russians if there had been more imagination." Ley's articles on Goddard prompted G. Edward Pendray to write an angry letter claiming that Ley was misrepresenting the secrecy of Goddard. This started a brief and nasty back-and-forth with the editors of the *Chicago Sun-Times*. In a letter to the editor, Ley claimed never to have read Goddard's book before arriving in the United States. Pendray lashed back, claiming that Ley was the chief architect of a German conspiracy to diminish the contributions and legacy of Goddard. In spite of Ley's newfound efforts to promote indigenous American rocketry, Pendray continued to weave together conspiracy theories that implied duplicity on the part of the Germans.[34]

Meanwhile, in "Space Science Pleas Ignored," Ley made the case directly by lamenting the fact that American scientists and engineers faced ridicule and public doubt. "America snubbed her scientists," the headline argued. In general, scientists "were told to take their science fiction plots home with them." Then, according to Ley, "we lost the largest propaganda battle of the cold war," because of the "bad impression" of using a military rocket to launch a satellite. The decision to use Vanguard instead of Orbiter was a fatal mistake. He argued that "we would probably have an unmanned rocket on the moon right now," if they had given von Braun the green light.[35]

Additionally, Ley assaulted official explanations: "We have suffered a totally unnecessary loss of international prestige. . . . We have lost a major propaganda battle." Whereas Ley downplayed this propaganda victory in early October, he now stressed its global importance. Ley educated Americans about Soviet rockets and space superiority. Regarding Russian claims, Ley usually added a cautious word on Russian announcements, which indicated a consistent distrust of Soviet claims. He reluctantly credited the Russians with many accomplishments, even the launch of a "liquid fuel research rocket" between 1935 and 1937. He added, "The mystery is not that they did it but that they kept it a secret. Otherwise the Russians have been rather ready to brag about all of their 'firsts,' even to the extent of claiming a few which certainly were not theirs."[36]

In his view the Russians were leading the way into space simply because they "just went to work" earlier than the Americans. For example, Ley commented on a radio interview in which someone said, "I bet the Russians can't just go to a library and read all about the latest rockets." Ley investigated the Russian literature. To his surprise, he found that several notable German and French books had been translated into Russian. In contrast, there were no English equivalents. Ley announced, "We ought to get busy and do some translating ourselves. Other people have ideas, too!" Americans should have been less complacent and more internationally minded.[37]

Increasingly, Ley also discussed the military potential of spaceflight technologies. For example, in "Conquest of Space Vital for Nation," Ley argued that the Russians "could prevent us from using space if they got there first." Space superiority meant that the Soviet Union "could do as it pleased over enemy territory." Ley pleaded, "The immediate and urgent purpose is to establish space superiority." The space station would give the United States a strategic advantage over the Soviet Union. It would also provide a base that might launch missiles of its own. Ley concluded, "We cannot afford not to have space superiority. Fortunately the space station also offers the promise of new discoveries and, since discoveries very often pay off commercially, it is rather likely that the money spent for space superiority will actually be an investment." Spaceflight would either pay for itself or result in enormous dividends.[38]

Ley continued to speak about cultural lag and missile gaps during public appearances. For example, he told an audience at Los Angeles

State College, "We are lagging behind Russia." If the Russians soon accomplished more victories, "it is to be expected." The *Los Angeles Times* reported on Ley's perspective: "The inconsistency of a powerful country, one of whose citizens built and launched the world's first liquid fuel rocket, being nominally two years behind Russia appalls him." The newspaper quoted Ley as saying, "The reason Russia is two years ahead of us is because of a difference of approach. We waited until a hydrogen warhead had been reduced in size sufficiently before we went ahead with a missile to carry it." Ley added, "Unfortunately, the United States is perpetually cursed with official thriftiness. . . . Naturally, I favor democracy, but sometimes it is too flexible. We need longer terms for our office-holders to give a continuity of effort." He allegedly stated, "Can we catch up with the Russians? Nobody ever said we can't, but nobody ever said it will be easy."[39]

He made similar statements during other lectures. Ley also answered the public's central question, "How could the Russians . . . ?" Ley summarized his "simple" answer: "If you round up, say, one million people anywhere this 'sample' will contain so-and-so many geniuses. . . . Add to this a government which has a goal and purpose . . . the bright boys and girls will get a chance to show just how bright they are and what they can do." Ley then summarized the reaction of the crowd. A person usually responded, "But . . . but we have never heard of any Russian geniuses." Ley also wrote the foreword in a book about Russian rockets. He stated: "I hope this information will contribute to ending the nice cozy nap from which so many do not wish to be awakened."[40]

Despite his anxieties about the Soviet Union, Ley saw positive signs by late 1958. For example, in the *Rotarian*, Ley reassured the public that a Soviet flag on the moon had no legal standing. In other articles he offered readers some reasons to hope that the Russians might be falling behind, while the United States lessened the missile gap. With the creation of NASA, Ley predicted a new rocket that "will resemble the present Vanguard in name only." Ley also favorably contrasted American and Soviet developments, by arguing, "American scientists will not send up a man until they're sure they can get him back safely—while the Russians probably won't be too concerned over the fate of the man."[41]

Models and Displays

Throughout 1958 and 1959 Ley's tactics were not confined to television appearances, public lectures, and print media. In early 1958 he helped to organize a "space exposition" at New York department store Abraham & Straus. The vice president of the company told a reporter, "We think this is the biggest, most comprehensive public space show in the country—I don't know about the world." The *New York Times* described a mesmerizing scene of "rockets, models of rockets, rocket engines, space suits, unborn space ships, and similar intergalactic gadgets." The exposition also included a cutaway of the combustion chamber of a V-2 rocket, a "mock-up" of a "Navy Stratolab balloon gondola," and a "scale model of the Jupiter-C rocket." Obviously the space exposition capitalized on the successful launch of *Explorer I*. When the show opened, "scores of school children" entered the store with their teachers and parents. This exhibition celebrated post-*Sputnik* American accomplishments. Most likely the event excited the children and reassured the adults.[42]

Ley also brought space models into the homes of American children. In early 1959 he entered into association with toy maker Monogram to serve as a consultant on a line of space models. Monogram announced this association with great optimism about the future. Its newsletter read, "Monogram has taken a GIANT STEP forward in our association with Willy Ley. . . . Impatient customers want 'authentic' space models NOW!" The newsletter also boasted of Ley: "No idle dreamer, he, but a man with deep conviction, one of the few scientists who, throughout the years, has been expounding the philosophy of space and space travel." As such, Monogram's new models would offer revolutionary glimpses into the future, when "there will be an entirely new concept of life and of living."[43]

Monogram then showcased four authentic space models, as designed by Ley. The promotion argued, "They will feed the thirst that is being created hourly in the minds of millions of people, young and old, all over the world." The newsletter encouraged its distributors to jump onboard: "All of us in this business have one of the greatest selling opportunities." Monogram advertised its models in newspapers and magazines as well as on television and radio. "The entire nation," Monogram announced, "is being told about Willy Ley and his connection with Monogram Models." In a later newsletter Monogram claimed that a day scarcely went by without a "space conscious America" seeing, hearing, or reading about their models.[44]

The announcement quoted a retailer who argued, "People are usually very interested in the future and the Willy Ley name seems to make them unusually real." A public relations man also claimed: "Mr. Ley is quite a personality, and I might add, quite a celebrity. Because of the nature of his models and their value as disseminators of information and a better understanding of what is to come in the space age, many publicity channels have been opened to us." An advertiser agreed by stating, "They represent a new era in human existence and have a tremendous educational value for adults and youngsters alike."[45]

As this publicity campaign ensued, Ley made many appearances on radio and television, and he attended the industry's annual trade show. He enthusiastically promoted the educational value of his space models. Ley claimed, "The best way to learn is by doing. . . . To-day's children don't have to unlearn things before they can understand what we're talking about. A child accepts an explanation of space travel, an adult argues about it." Ley also used his connection with the *Chicago Sun-Times* to promote the models. The supplemental magazine, *Sunday Midwest*, ran a full-page story that took readers on a trip to the moon in a future vehicle based on Ley's designs. Monogram soon published a "Space Age News Letter," edited by Ley.[46]

By late 1959 Monogram proclaimed success in their advertising campaign. Inside the October flyer, Ley reflected, "Perhaps slow at first to catch on . . . I am now glad to report that the models have steadily grown in popularity, and tremendous numbers of them are being sold." He attributed the popularity of the models to their accuracy of design. He claimed, "I designed them with the full force of facts and research findings at my command and proceeded as I would if I were designing a real space vehicle, for actual space travel." Ley argued, "These kits represent delightful fun beyond comparison; provide the means to fascinating space development study; give a better understanding of space travel than you can obtain in any other way."

By constructing the space models, children would participate in the imminent conquest of space.[47]

Disney in the Classroom

In addition to bringing space models into the homes of children, Ley spent much of 1959 bringing Disney's earlier lessons into the classrooms and libraries of public schools. His small and lavishly illustrated books were based on Disney's spaceflight programs and "adapted for school use." Ley's first adaptation was *Man in Space* (1959), a forty-eight-page book. It contained many of the familiar themes of his Sugar Jets books. It also relentlessly promoted von Braun and his designs. While dramatic illustrations depicted the first manned flights, as well as a "space walk," Ley portrayed the events as routine engineering. Regarding the problem of re-entry and landing, he wrote: "When the ship is within the earth's atmosphere it will be flown just like an airplane. It is now simple to aim for the base and its runway. . . . Then touchdown. The first orbital flight is over." This tactic of combining dramatic descriptions and simple explanations of routine technological feats had become a staple of Ley's texts. He almost always ended his juvenile books with a nonchalant description of everyday spaceflight.[48]

His other contributions are interesting. For example, *Mars and Beyond: A Tomorrowland Adventure* taught young readers about the future exploration of the red planet. Ley wrote, "Some fifteen years in the future, powerful telescopes will be able to spot a strange structure orbiting around the earth: the first Mars ship." The book also made the case for a large expedition, similar to the ambitious designs of von Braun that had aired on television. Ley wrote, "A fleet of ships would be best for an expedition to Mars." Unlike the more modest trip in *The Exploration of Mars*, Ley describes a massive convoy of umbrella-like space ships that function on atomic reactors. Once they arrive at Mars, they launch "landing boats," which parachute down to the surface as reaction rockets slow their descent. This adventure was the only way that scientists would solve the mysteries of Mars. Human beings had to get there and look around. The book concluded with a glimpse into the future, when Americans had colonized Mars. Colonization was a matter of routine. People would simply "leave the caves and build domed cities on the open plains." It would be an easy and routine affair.[49]

Looking in from the Outside

During the earlier *Night-Beat* interview in 1957, Mike Wallace prodded Ley: "Willy, as a science reporter, how much do you yourself really know about missiles, in the face of the military security which is put about them?" Ley answered, "Eh . . . that depends on what part you look at. If you look at the question of propulsion, not very much is secret, as far as I know. If you look at guidance, practically everything is secret." Ley argued that much of the secrecy had been "overdone." He gave an example of reading recently declassified documents. He joked, "No amount of questioning could make me think of a reason why they should have ever been classified in the first place." He felt confident of his knowledge as an outsider.[50]

Wallace then asked one last question: "Seriously, or more seriously, we should say on this one: Up to the time you left Berlin, you were among the most promising rocket research scientists. Today, you're on the outside. You're writing about scientists. You worked with them. Do you ever feel that you are missing out on the really creative work?" Apparently, none of Ley's activities counted as the "really creative work." Ley understood the sentiment behind the question: "Yes," Ley admitted, "but I have a good substitute . . . if I were engaged in actual research work, I would be engaged in one small field. As a science writer, looking in from the outside, I can watch simultaneously a dozen interesting fields." Ley's tone of voice indicated his firm conviction that the outsider could be an expert. In fact, the outsider could become an interdisciplinary guide, seeing from a public vantage point. Conversely, the specialists could be lost in isolation. There were additional obstacles of military secrecy and inter-service rivalries.

At the end of their encounter Wallace thanked Ley for the interview, before speaking to the audience: "Somewhere between us and the scientific mind stands Willy Ley, always curious, and always ready to translate the complex into clear, understandable, but still highly adventurous terms. We hope that when the first rocket ship reaches the Moon . . . Willy Ley will be aboard, and we'll bring back to *Night-Beat*, as a guest: a genuine, bona fide moon man."

Ley would have taken the ride. Despite his lack of engineering skills, his informal education, and his disconnect from centers of research and development, Ley could still claim to be a rocket expert. The general public would have accepted his right of passage into the cosmos. This acceptance teaches us much about the Space Age and the role of popular science. If historians wish to understand both the public hopes and fears throughout 1957 and 1958, then it is crucial to recognize the powerful influence of a science writer who worked outside the walls of an institution. Willy Ley had little inside knowledge of American missile programs and the true nature of the Space Race. During the late 1940s he had desperately struggled to insert himself into the field of rocket engineering. He criticized the reliance on von Braun. Yet by 1957 Ley was seen as one of America's leading rocket experts. He was also a media insider, who could promote von Braun as a brilliant engineer. Journalists and citizens turned to Ley for answers and predictions. Both before and after the *Sputnik* moment, he took every opportunity to excite his audiences, whether they were attendees of an annual meeting or curious children reading with flashlights in bed. His works document not only the popular excitement and expectations of an American future in space but also the fears and anxieties. Ley influenced the perspectives of millions of Americans who embraced a vision of the future with its corresponding and optimistic expectations. More than many other figures, he shaped those expectations. Although he lacked the hero status of von Braun in 1958, he was still incredibly influential.

By 1959 Ley and von Braun had formed an alliance. They needed each other for crucial tasks. Together they fostered much public support. While von Braun became an important insider, Ley looked in from the outside. He was disconnected from important decisions, closed-door meetings, and inter-service rivalries. He influenced fellow citizens. In his capacity as a public expert, Ley gained a reputation as reliable and trustworthy. One reporter described him with these words: "Willy Ley is indeed an unusual person—a refugee from Hitler's

Germany who now works in an unofficial capacity for the United States Government as disseminator of information and a vital force for national and international understanding."[51]

Although he lacked von Braun's hero status, Ley shared the stage in an influential way. In fact, both men sat down together as equals in 1959 to record their memories for a Vox LP. They reminisced about the early days in Berlin, before moving onto to more recent topics. They ended their talk with hopeful comments about extraterrestrial life. Together they had done so much to initiate the conquest of space. During the long interview, both Ley and von Braun often interjected, "Yes . . . I remember."[52]

Scholarly Twilight

Willy Ley spent the last nine years of his life out of the limelight. He let von Braun and others do most of the space-related work and publicity. In some ways this period could be characterized as an era of fading celebrity status and misplaced priorities. For example, science fiction historian Sam Moskowitz described Ley's final years as "Losing the Last One." Moskowitz argued that Ley "could no longer maintain a dominant position as a popularizer of space knowledge." Moskowitz also argued, "He was running faster and faster to stay in the same place." Allegedly, Ley struggled to stay afloat from a combination of product endorsements, royalty checks, lecture fees, and various odd jobs. To make matters worse, according to Moskowitz, Ley's efforts were largely wasted on semi-scholarly books that failed to attract large audiences. He also struggled to keep his bestselling *Rockets* up to date.[1]

For the most part Ley stopped writing for newspapers and magazines, with the exception of his *Galaxy* articles. In the scholarly twilight of his life, he devoted himself to the histories of astronomy, zoology, and natural history. For Ley, this historical focus made sense. By the early 1960s the Space Age flourished. NASA would soon put an American into space. While a cast of astronauts became celebrated American heroes, Ley devoted himself to his first true passions.

This period should not be labeled as "Losing the Last One." Although Ley was working on many competing side projects, he displayed an enormous focus on two large histories of science. First, his *Watchers of the Skies: An Informal History of Astronomy from Babylon to the Space Age* (1963) was a massive opus on the history of astronomy. Second, his *Dawn of Zoology* (1968) presented his most comprehensive account of the "pre-history" of modern science. By analyzing these two books in detail, we can understand how his vision of science and its past related to other developments, particularly the professionalization of the history of science as an academic field. If there is anything inherently sad about this last decade of his life, it is the fact that younger historians of science ignored his contributions. His earlier ventures into the history of science were called "scholarly," yet his new ventures (which were overtly more scholarly) were seen as popular and irrelevant. His histories of science became increasingly out of step with the intellectual scene. His unique blend of romantic naturalism and modernism would not sit well with a new generation, which became far more critical of science and technology.[2]

Continuities and Conquests

During the 1960s Ley did not fade completely into obscurity. He continued to speak on radio and television, in between or during lecture tours. His "Conquest of Space" lecture became so influential that H. W. Wilson Company included it in *Representative American Speeches: 1960–1961*. Simultaneously, Ley still wrote about rockets and space travel, mostly in niche publications. For example, he contributed several articles to the newly founded and struggling magazine *Space World*. His most interesting pieces dealt with the military potentials of spaceflight technologies in the context of the Cold War.[3]

In "A Fortress in Space" Ley argued, "it is obvious that there has to be military preparedness in space for safety's sake." After ruminating on the space station's tactics, Ley stressed, "We cannot neglect space defense." A newsletter also warned of the consequences of Soviet superiority: "Near-future space feats of the Soviets may include hurling a 50-megaton H-bomb to explode on the moon." Ley concluded, "All would be 'ballistic blackmail' to . . . panic America at Russia's 'overwhelming space-power'—and that statement might then be true, jittery experts warn." Ley also warned of a future "Victory gap" in the coming space war, because "Washington is keeping military men out of space policy decisions . . . [while] Moscow is loading its top leadership . . . with uniformed experts." In spite of his earlier perceptions of the totalitarian control of science, he pleaded for total mobilization and central oversight. Like other scientific intellectuals, Ley displayed little awareness of the intense contradictions. He did not reflect on the incorrect assumptions about totalitarianism and science. Americans simply had to adapt to changing political and technological realities.[4]

Ley wrote other space-related articles for *Galaxy*. One of his most noteworthy articles was titled: "Are We Going to Build a Space Station?" Ley reassured readers that the scientific and engineering goals had not been lost amid current debates. Other articles included "The End of the Jet Age," "Sounding Rockets and Geoprobes," and "Anyone Else for Space?" When Ley was not updating readers on recent developments, he continued to explore the prehistory of American and Soviet rocketry. Yet in *Galaxy* his spaceflight pieces became less frequent. He obviously enjoyed writing different articles like "A Century of Fossil Men," and "The Rarest Animals." These writings reflected an agenda that was broader than rockets, missiles, and space travel.[5]

Science as the Humanities

In the September 1960 issue of the *Instructor*, Ley argued: "It's not science *Versus* the humanities." He wrote, "Among the things I do not understand . . . is making a distinction between 'science' and 'the humanities' with resulting discussions about which one should be emphasized." Ley pleaded: "Such talk has no meaning whatsoever, for if 'the humanities' means the elements of our culture, then certainly science is part of the humanities." In other words, science was a human activity, which could not be divorced from culture. Science reflected human values, beliefs, and aspirations. To talk of "two cultures," as chemist/novelist C. P. Snow had done, was simply nonsensical and misleading.[6]

As Snow's *The Two Cultures* (1959) created controversy, Ley paraphrased an editor who valued the distinction, since a subject like history "can be 'really taught' while the pupil must take the results of scientific research at face value." In Ley's view this perspective was absurd. The opposite was far more accurate: "The teacher has been taught in college that the greatest epic of classical antiquity is the Iliad. . . . She passes this knowledge on to her twelfth-graders. . . . The students must accept these statements, for the most they can do is to read samples of their work in translation." "Even then," Ley argued, "they still have to take the teacher's word at face value." Conversely, science was far more open to independent verification. Ley wrote, "Compared to what the pupil would encounter in checking the claims of 'the humanities,' there are virtually no problems in checking the statements of elementary science." Every child, Ley claimed, could easily confirm the validity of the Pythagorean Theorem or the weights of lead and iron. This verifiability, as well as the social importance of knowledge, put science at the top of the humanities. Ley concluded, "Today's surroundings are mostly 'science,' and I hold that the purpose of all teaching is to create an understanding of our surroundings." Science was "not beyond understanding." As Ley argued in a draft manuscript for a book, "The things that are not beyond understanding are the laws of Nature that are around us every minute of our lives from birth to death."[7]

In a different article Ley outlined the broader agenda. He asked, "Do we want to make everybody a scientist? Do we want to make everybody a humanist?" He answered, "We couldn't do that. But, what we could, and should, do is to produce in the next generation and in ourselves by application . . . the equivalent of what you call the music-lover or the art-lover." The educator should guide his students "in the direction of being lovers of science." The educator should train "an understanding public." Ley added: "What we must produce is a scientific equivalent of that [music appreciation], not a young man or young woman who learns science to practice it later but who learns science . . . to be able to follow what the scientists are doing and not say, 'I don't understand this.'" Ley also argued, "The effort should be directed, I feel, in producing an audience for the scientists."[8]

Like other science writers, Ley used case studies in the history of science to advance this agenda. The history of science would serve as more than a simple interface between two cultures. It would demolish the distinction, while celebrating the bold adventures of the past, the progress of Western civilization, and the universality of human knowledge.

The Conquest of the Earth

In 1962 Ley wrote *The Poles*, a Life Nature Library book that combined breathtaking photographs with dramatic text. Ley titled his chapters with phrases like "The Cold, Far Frontiers," and "The Great Antarctic Laboratory." In style and format the book resembled Ley's earlier *The Conquest of Space*, albeit focused on the exploration of "the extremes of the earth." Key themes included the mysteries of the unknown, the sagas of brave explorers, and the vast scientific and economic opportunities of human conquest of the North.[9]

Ley began the book with a sweeping prediction: "The poles have long been a challenge, remote, and forbidding, to man's sense of venture and curiosity. Today, in the sweep of technological revolution, the Arctic has become a highway, and tomorrow both the polar regions may be exploited for food, minerals and other materials needed to support man's steadily expanding population." Ley showcased the scientific wonders of the poles, before devoting two chapters to the history of exploration and the "questing human spirit." Ley recounted the dramatic, exciting, and often tragic tales of human endurance in unforgiving frontiers. The chapter on arctic exploration concluded with these words: "Man has indeed become master of the Arctic, but he has many earlier men to thank for it." Ley glorified the fearless adventurers.[10]

Ley also examined the "odd polar animals," such as the "perky, peculiar penguins," the "ugly elephant seal," the "wondrous walrus," and the narwhal, which Ley called "the unicorn of the arctic." The book then focused on the strange and curious people who inhabited certain regions. Ley discussed the "primitive peoples" who are "preservers of Stone Age life." He admired the "arctic aborigines," while he simultaneously fantasized about modernization. Ley expressed fascination with a paradoxical reality: "Their survival is a supreme example of human adaptability. Their happiness, springing from a philosophy and way of life that civilized people have long abandoned is a triumph of the human spirit." Ley cautioned readers by stating that "civilization is not to be confused with culture—the complex of beliefs, customs, institutions, tools and techniques by which a society lives—and the arctic peoples developed some of the world's most ingenious and interesting cultures."[11]

Despite this admiration for indigenous cultures, Ley anticipated few negative consequences to the civilizing process that would accompany modernization. In fact, he made the case for inevitable racial harmony: "Whites and natives are mixing and changing. . . . Eskimos and Indians have become American military pilots and riflemen . . . Lapps in Norway publish their own newspaper, and the Athabaskans at Fort Yukon have formed a jazz band." Arguably this passage illustrated the complexities of modernization, which did not simply bring civilization to the frontier. Rather, civilization (and Western culture) adapted. Ley argued, "By now, the mixing of aborigines and immigrants is almost complete." This "new breed" will have "adapted to the strenuous requirements of the Arctic, and the struggle for existence has toughened their bodies and developed their minds." Ley added, "They are a hardy, independent people with a remarkable spiritual and material culture, and they are certain to play major roles in the coming arctic boom."[12]

The remainder of the book celebrated the inevitable boom. Ley predicted: "World trade and travel are taking to the skies, and the north is their great short cut. World power has passed to the continent-sized nations rimming the Arctic, and the north is the frontier they must man and maintain." Ley viewed these trends in the context of the Cold War: "Because so much Soviet land and resources lie so far north, the Russians have taken the lead in opening up their northlands." Fortunately, Canada and the United States were making up lost ground. Ownership of the north became an issue of vital importance, because "Population, technology, military needs . . . are thrusting civilization farther north." Ley was quite optimistic about the untapped resources of the "Great Tomorrow Land." One illustration presented a domed city of the future that protected its inhabitants from the cold and desolate landscape.[13]

In Ley's perspective, humans would transform a barren and hostile environment into a world of tomorrow. The human spirit would prevail over the challenges of nature. Conquest was inevitable and glorious.

Watchers of the Skies (1963)

Throughout the early 1960s Ley worked on his opus on the history of astronomy: *Watchers of the Skies: An Informal History of Astronomy from Babylon to the Space Age* (1963). He stated his ambitions: "Astronomy, all historians are agreed, is the oldest of the sciences, with the automatic result that its history is not only of great length but of extraordinary complexity." Earlier attempts to present a comprehensive history resulted in failures. Ley argued, "They tried to look 'historical' by mentioning a few names and dates of the past but were far from historical in that they did not even discuss the thoughts, correct or mistaken, of the people mentioned." Ley aimed to produce a definitive history that avoided these pitfalls. He clarified his methodology: "One should let the people who made that history speak for themselves." Ley not only sought to present historical actors on their own terms, but he also sought to present a unified portrait of astronomy. The history of astronomy desperately needed that sense of unity.[14]

Additionally, Ley made few apologies for what the text excluded. Any discussion of cosmogony belonged in history of natural philosophy rather than astronomy. Likewise, a discussion of Chinese constellations belonged "in a book on Chinese culture." "Similarly," he argued, "the star myths which were developed by many 'primitive' peoples have been disregarded." His massive and unified history of astronomy would be inseparable from the history of Western thought and culture. His attempt to unify astronomy resembled a Eurocentric umbrella that celebrated Western progress.[15]

In part one, "A Science Grows Up," Ley narrated an entertaining and educational history of stargazing from ancient times to "The Celestial Century" that followed Sir Isaac Newton's *Principia*. Ley discussed the "proto-astronomy" of early religions. For proto-astronomy to transition into a science, three conditions had to be met. The sky had to be clear with some degree of regularity. A certain number of people needed leisure time to make observations. Most important, they needed the means of recording those observations. All three conditions came together in ancient Mesopotamia, with Persian and Babylonian "priest-astronomers." Yet the ancient astronomers did not possess a true sense of wonder, according to Ley. Instead, they utilized observations for time-keeping and ceremonial purposes. He also attributed a "future pattern of astrological beliefs" to the Babylonians. Ley then focused on the Chaldeans, who conducted "observation, not distracted by a search for omens." Despite their clear-minded approach and careful observations, the Chaldeans were still stunted by their religious worldview. Ley added, "Their calculations were a form of worship." Consequently, their astronomy "petered out." Ley also devoted many pages to debunking nonsensical claims about ancient Egyptian astronomical knowledge. After this "excursion into silliness," Ley turned to ancient Jewish scholars, who "rejected the study of the sky as both 'foreign' and 'godless.'" Due to their religious beliefs, astronomy "could not develop among the Jews." Science had to overcome superstition and mysticism.[16]

In this perspective the ancient astronomers were blinded by cultural and religious beliefs. The Greeks served as the main exception. Their philosophical debates took them in the right direction. Plato's upstart disciple Aristotle tried "like Alexander von Humboldt . . . to know everything." Rather than defer to a sacred text or legend, Aristotle's goal was "to produce a complete picture of the universe." In a feat of independence and bravery, Aristotle broke with Plato's universe of ideas. For Aristotle, "the world of visible phenomena was the real world." Accordingly, Aristotle was the first real scientist urging observation and even some degree of experimentation. Unfortunately, his greatest virtue was also his greatest flaw. "Aristotle's fundamental mistake," Ley wrote, "was to underestimate the magnitude of the task [of knowing] and to conclude that he personally knew enough to draw conclusions."[17]

Ley then discussed the contributions of other Greek philosophers. The discussion of Greek astronomy also contained a long evaluation of

Ptolemy and his legacy. In Ley's perspective Ptolemy deserved praise for only two things: he labored diligently, and he preserved the work of Hipparchus. Otherwise "he showed himself to be as reactionary as possible." Ptolemy deferred to the authority of Aristotle, while he remained blinded to the truths around him. "Why didn't he open his eyes?" Ley asked. Instead of evaluating the evidence and experimenting, Ptolemy "strained all resources of rhetoric and argumentation." Ley added, "It is quite possible that he was a pure theorist."[18]

What followed can be summarized as a long period in which astronomy slowly lingered in darkness. While Ley praised the Arabic efforts to preserve Greek texts, he saw little cultural exchange. The history of astronomy suffered through an "interregnum." Ley explained: "No real progress was made. . . . One might even argue that Ptolemy's dogmatic adherence to the view of that earlier era contributed heavily to the sterility of the interregnum." Instead of flourishing cultural exchange and the spread of new ideas, the medieval period is portrayed as one of deep stagnation and deference to established authorities. Ley argued: "The later Middle Ages indulged in a kind of Aristotle cult." Overall, the medieval scene was a "long and naturally sterile period."[19]

The reign of authoritarian dogma ended during the Renaissance, when brave and bold adventurers challenged established thought. Their ties to broader movements were obvious to Ley. He wrote: "On October 31, 1517, one Martin Luther had nailed ninety-five theses on the heavy oaken doors of the church of Wittenberg, Germany. And on February 19, 1473, Nicolaus Copernicus had been born." Ley thus implied a direct connection between the Protestant Reformation and the Scientific Revolution. Simultaneously, Ley lamented the fact that Copernicus was still a product of his time. Regarding the *Commentariolus*, "Those opening pages . . . sound so 'modern' that a reader of today who proceeds further suddenly feels something quite close to disappointment. This 'modern' concept is darkened by the persisting epicycles." Ley continued: "Copernicus has epicycles running on epicycles."[20]

Nevertheless, the revolutionary spirit and system of Copernicus would inspire other bold and daring thinkers to nail their theses to the proverbial doors of science. Much impetus from this iconoclastic spirit came more directly from the invention of the telescope and the "inevitable collapse of whatever remained of Aristotle's philosophy." In some ways the telescope was unnecessary due to "Tycho's Star," a supernova "which everybody had seen." The notion of an immutable cosmos had collapsed overnight. Resistance to new ideas continued, particularly with Tycho Brahe's attempt to "save the phenomena." Nevertheless, the tide of change could not be stopped. Like Aristotle's break with Plato, Johannes Kepler broke with Tycho Brahe. While the earlier systems had been "darkened" by persisting nonsense, Kepler saw the light. As such, "he became the intellectual successor and 'completer' of Nicolaus Copernicus."[21]

Medieval superstitions still haunted Kepler, particularly when the public demanded astrological explanations. Ley presented Kepler as a public educator who walked a tightrope between science and sensationalism. Kepler "simply repeated what others had said: that the world would burn up, that all Europeans would move to America, that the Realm of the Turk would be destroyed . . . and so forth." Regarding astrology, "Kepler—who after all was a child of his time—generally believed in the idea of astrology, disbelieved its rules, but cast horoscopes according to these rules because he was paid for them." In spite of this tendency to dabble in mystical or pseudoscientific enterprises, Kepler ventured forth, fearlessly concluding that perfect, circular motion had been the reigning myth of astronomy. Kepler debunked the myths and countered the skepticism and resistance of even his close friends.[22]

Galileo Galilei carried the torch further. Viewing the universe with new eyes, Galileo discovered things that were "easy to test experimentally, and Galileo (this is where he differed from most of his contemporaries) did test." Ley then explored the history and myths surrounding Galileo's famous experiment atop the Tower of Pisa. According to Ley, Galileo "was just conducting an experiment which he probably knew had been performed before." His real contribution to the astronomical scene surrounded his detailed observations of the distant stars and the moons of Jupiter, as reported in the *Starry Messenger*. The implications of Galileo's observations were obvious, even to Kepler, who doubted some of the conclusions. Yet Kepler "went further," arguing, "Produce ships and sails which can be used in the air of the sky. Then you'll also find men to man them, men not afraid of the vast emptiness of space."[23]

The remainder of the chapter recounted Galileo's many public battles against detractors and philosophers. Implicit in Ley's portrayal is an admiration of a fearless and anti-authoritarian debunker, who often took his case directly to an educated audience. Galileo faced powerful opponents who stood "united only in the beliefs that there was no need for new facts, that the facts were already established, and that they could be found in Scripture." As a man of science, Galileo braved "an all-out attack" from the authoritarians and their pseudoscience. The "main blow" came from literal interpreters of scripture. Powerful traditionalists stood in the shadows, waiting for an opportune moment to strike the scientist down in "the battle about the structure of the solar system."[24]

After Galileo was "officially silenced" in 1616, he "could not keep quiet for very long." He was too convinced of the truth, and he was too determined to bring that truth to the people. He could not stand idly by while dogmatic beliefs and nonsensical philosophical systems reigned. He had a duty to expose the darkness to light. Eventually Galileo published his *Dialogo*, making the case for Copernicanism. Ley viewed it as a work of popular science that "was an immediate success with many people." According to Ley, "the literary public was delighted, [although] quite a number of Church Fathers were aghast." Galileo's critics were also horrified that the book was written in Italian "so that anybody could read it." Ley clearly appreciated what Galileo tried to accomplish, and he expressed confusion at Urban VIII's negative and irrational reaction.[25]

Unfortunately, anti-scientific forces struck back at the scientist. As a devout Catholic, Galileo "had to submit to authority in some manner." Ley reiterated his support for historian Giorgio de Santillana's conviction that the "thundering theological persecution" disguised the real motives of Galileo's enemies. Ley quoted: "They could not very conveniently broadcast the real motives, which were that Galileo had taken to writing in Italian and that he had made them look foolish, or that the political meaning of it was that the Jesuits had evened up a score with the Dominicans by way of the new game of cosmological football." Ley even defended the legitimacy of a popular legend, in which Galileo defiantly said, "And yet, it moves." Readers are left with the impression of Galileo as a man lost in an age of backwardness.[26]

With post-Galileo astronomers, Ley had a difficult time retaining the chronological narrative. In fact, much of the structure of the book became disjointed. Nevertheless, the remainder of *Watchers of the Skies* continued several themes. Astronomers are depicted as bold theorists, telescopic explorers, and fearless adventurers standing up to dogma, while writing for the educated public. Quite often mysterious comets or astronomical hoaxes panicked people. Ley took aim at certain popularizers who exploited the fears. He also explored the ways in which astronomers even contributed to sensational reports. Ley concluded the book with an interesting discussion of the "Search for Other Civilizations." This epilogue brought a central narrative back into the text as Ley excited readers with the possibilities of extraterrestrial life. Astronauts would serve as the next generation of brave astronomers. A ship would soon depart.[27]

Ley's opus on the history of astronomy arguably failed to bring unity to the field. The scholarly format of the text also convinced science fiction historian Moskowitz that the book was a "crushing disappointment." Moskowitz further claimed that although the book went through five immediate printings, "Willy never got much money out of it." It had been a monumental project. Yet "three years had passed. . . . A dozen more profitable projects had been turned down or postponed." Ley had wasted time, Moskowitz argued.[28]

Moskowitz's tone is somewhat insulting, given that Ley was not attempting to write a bestseller. It also ascribes Ley's motivations to profit, as if his three-year devotion to the history of astronomy was a financial gamble. This perspective ignores Ley's true motivations for devoting himself to a project that contributed to the history of science. It also ignores the fact that the book was successful. Five back-to-back printings indicated that it was selling quite nicely. In fact the November 1963 issue of *Book-of-the-Month Club News* praised the book: "There is something here for everybody interested in the progress of human knowledge." The reviewer also stated, "Mr. Ley brings in an astonishing wealth of unexpected information." Owen Gingerich of the Smithsonian Astrophysical Observatory also complimented "the vast amount of scholarly researching and digging that has been spun into this imminently readable account."[29]

The book was also well received by several historians of science. None other than I. Bernard Cohen, former editor of *Isis* and recent president of the History of Science Society, praised the book in the *New York Times*. Cohen stated, "I know of no other source for such authoritative and easily available information on the history of each of the planets as this new book." Cohen concluded: "Many delighted readers will be grateful to Willy Ley for having provided them with so stirring an account that combines the virtues of readability and an awareness of the latest historical and scientific research." Other reviewers praised Ley's nuance as a historian, while concluding: "In the face of the universe, all men have been brothers, scientifically speaking at least."[30]

Not all reviewers were positive. Historian of science C. Doris Hellman remarked in *Isis*, "Whereas the book is recommended for the layman, it is not a reference book nor a book for the historian of science. The spritely, sometimes flippant style may grate on the student who won't like the way people are introduced. . . . Nor will the student feel happy with the numerous minor inaccuracies and misprints." At other times, "the scholar will wince."[31]

A Changing Scene

Ley's *Watchers* did not reference a recently published book: Thomas S. Kuhn's *The Copernican Revolution* (1957). According to a later inventory of Ley's personal library, he owned the book, although it is unclear when he obtained a copy. It is easy to imagine Ley reading the text with a critical eye. Kuhn claimed, "This book repeatedly violates the institutionalized boundaries which separate the audience for 'science' from the audience for 'history' or 'philosophy.'" Ley would have cringed at this statement as an implicit recognition of "two cultures." Kuhn also claimed, "Except in occasional monographs the combination of science and intellectual history is an unusual one." How could the works of Sarton, Butterfield, and others be labeled as specialized studies, when they often served as general and intellectual histories of science? Kuhn added: "Scientific concepts are ideas. . . . They have seldom been treated that way." Ley would have been baffled that a scholar could so easily dismiss a generation of intellectuals who had popularized the history of science as a fascinating mixture of revolutionary ideas and upstaged beliefs.[32]

Kuhn's most highly praised book did not occupy Ley's bookshelves. In *The Structure of Scientific Revolutions* (1962) Kuhn argued that the history of science was mostly "a repository for . . . anecdote or chronology." Not only did earlier historians mislead readers, but they also failed to "to display the historical integrity of . . . science in its own time." They wrote "history backward." Kuhn argued that their "deprecation of historical fact is deeply, and probably functionally, ingrained in the ideology of the scientific profession." Kuhn challenged the prevailing attitude that scientific progress resided in the gradual accumulation of facts. Accordingly, historians needed to focus on the periods of revolutionary science, which led to new "paradigms." Science did not proceed in a straight line. Rather, cultural factors explained how science might alter its course. It took bold adventurers who collected the anomalies together and crossed into unfamiliar territory, where they fearlessly explored the alternatives, questioned the underlying assumptions, and generally expressed their anti-authoritarianism under the threat of a dominant scientific establishment.[33]

Although Kuhn offered a general model that quickly became both controversial and influential, many of these ideas were not new. Other writers and historians had used similar language for decades. Ley had long focused on conceptual shifts, scientific revolutions, and the influence of "irrational" elements in the scientific process. Indeed, readers could legitimately wonder how Kuhn expressed confidence in the novelty of these perspectives and why many of his contemporaries did not point to an earlier generation of popular writers who championed anti-authoritarian depictions of scientists, while presenting the history of science as quite messy, revolutionary, and culturally dependent. Certainly, there were Whiggish elements, yet to characterize their histories of science as uncritical celebrations of the linear accumulation of facts was incredibly inaccurate. To use the label of "Sartonism," in reference to founding father George Sarton, propped up an easily defeated straw man.

We can hint at a possible explanation for the changing scene by building upon the suggestions of other scholars. The task helps to explain the increasingly cool reaction to Ley and other science writers in the 1960s. Although certain academics (most notably Kuhn) continued to attract large readerships, a growing disdain for popular histories found expression in pages of *Isis*. For example, Robert E. Carlson reviewed L. Sprague de Camp's *The Heroic Age of American Invention* (1961). Carlson implied that such a popular book failed at key historical tasks: "Too often the author is overly concerned with the inventor's private life so that the economic and social environments in which these men lived receive only passing mention." De Camp's entertaining style seemed incompatible with the historian's demand for a more detailed examination of the broader socioeconomic forces.[34]

Other scholars criticized the efforts of science writers. For example, when Isaac Asimov moved from popularizing science to popularizing the history of science, Yale professor Frederic L. Holmes criticized his book as unscholarly and simply too brief. Holmes accused Asimov of "frequent oversimplifications or inaccuracies," as well as "distortions," "myths," and an overall ahistorical perspective. Asimov ventured further into the history of science in 1965, producing *Asimov's Biographical Encyclopedia of Science and Technology*. It was a massive undertaking that culminated in a 662-page reference guide to the history of science and technology. An *Isis* reviewer coolly noted that the book was obviously meant for "a general reader with broad interests, not a basis for scientific research in the history of science and technology."[35]

Willy Ley received a similar reaction to his translation of *Otto Hahn: A Scientific Autobiography* (1966). The book recounted the German chemist's struggles against a powerful and irrational regime. One of the book's central themes, in the words of a reviewer, surrounded the "evil political forces that nearly destroyed German science." Although the reviewer appreciated the central narrative of the text, he lamented the role of a science writer as translator and editor. During these years other historians cast a critical eye at scientific autobiographies as reliable historical sources. Autobiographies exemplified the problems of a scientist-turned-historian.[36]

Ley's related attempt to promote a compilation of Conrad Gessner's writings also earned criticism. Although the reviewer appreciated Ley's intentions, he wrote, "Willy Ley's New Introduction is a popularized account, pleasant to read but containing some factual errors and lacking any documentation." Ley's approach lacked scholarly diligence. In other cases historians began to identify certain writers as unwelcome outsiders to the field. This applied particularly to Arthur Koestler, whose 1959 *The Sleepwalkers* offended many historians of science. As a popular history of astronomy and cosmology, the book celebrated the irrational and unpredictable elements that led certain scientists to drift mindlessly in various directions. Koestler's portraits of individual scientists were often unflattering and quite provocative. He presented a rather totalitarian portrait of mainstream science, which hindered individual creativity and intellectual freedom. Modern science threatened romantic and spiritual values. To the dismay of many historians and scientists (including Ley), Koestler's book became a bestseller. Many historians denounced the book.[37]

In other cases a disdain for popular books reached fever pitch. For example, the Smithsonian's Nathan Reingold reviewed William Gilman's *Science: U.S.A.* (1965). Reingold argued that "too much of science writing is well characterized by the French term for popularization—vulgarization." Unfortunately, "historians cannot disregard the genre," because it contained useful clues from primary sources. In some ways it would be better, Reingold argued, if such a genre no longer existed. He announced, "Pity the poor future historian faced with a text lacking both bibliography and footnotes." He went on to make disparaging remarks about the intelligence of a science writer as compared to scholars and scientists. The popularizers were outsiders who could not be trusted. They vulgarized history.[38]

Attempts to popularize the history of science through film were also derided in the pages of *Isis*. David W. Chambers offered the following comments on the "Toulmin films," such as the short and educational *The Perception of Life* (1964). Chambers asked, "To whom, then, can such a film be recommended?" Because the "result is an anarchy of vaguely related images dazzling to the eye, [but] bewildering," Chambers wrote, "I offer the spirit, if not the letter of a phrase quoted in *The Perception of Life*: 'Investigations of this kind particularly recommend themselves to ladies.'" Popular science was relegated to an allegedly feminine sphere of science popularization.[39]

Other changing perspectives can be seen in the pages of *Isis*. A younger generation of historians turned a critical eye on a previous generation's political biases. For example, when Conway Zirkle extended his examination of the "death of a science" in the Soviet Union, he faced a critical reception. Scholar David Joravsky wrote, "Professor Zirkle is an angry man, and understandably so . . . he vents his smoldering anger on the founders of Marxism and on all those . . . who do not share his views on the biological determinants of history, on those who are 'dupes,' perhaps unwittingly, of 'Marxian biology.'" Joravsky highlighted "serious defects of his history," which too readily cast Soviet thinkers into authoritative camps of dogma. Joravsky would soon publish a monumental study on Marxism and science. Soon other historians reevaluated Soviet science.[40]

Other academic books critiqued the history of science as "pseudo-scholarly." Most notably, Joseph Agassi pleaded in *Towards a Historiography of Science*: "The study of the history of science is in a lamentable state: the literature of the field is often pseudo-scholarly and largely unreadable." Agassi went on to criticize the use of false dichotomies that separate "observation" from "superstition" as well as "fact" from "error." As the critique circulated, *Isis* reviewers increasingly used the label "nonscholarly" to identify popular books written by practicing scientists. In other instances a reviewer would express surprise: "Although the presentation is popular, the authors are abreast of the latest scholarship in the history of biology." The dichotomy between popular and scholarly writing became implicit.[41]

Meanwhile, the works of French philosopher/historian Alexandre Koyré began to influence historians. Koyré presented a more sympathetic portrait of religious radicals and mystical thinkers. A prominent scholar praised this new perspective: "He reminds us that sixteenth-century magic was a science." Koyré's style of reevaluating the importance of "pseudoscience" was not without precedent. This reevaluation had really begun a decade earlier, with Taylor S. Sherwood's *The Alchemists: Founders of Modern Chemistry* (1949). Yet several historians initially revolted against the alleged founding of a modern science in a mystical "pseudoscience." By the 1960s many scholars were far more receptive to such reevaluations. Historians of astrology also began to present more complex histories. The history of science was becoming more academic, self-reflexive, and critical. Historians became more concerned with disciplinary objectives rather than a popular crusade.[42]

Although Ley owned several of these books, he rarely incorporated these new perspectives in his own histories of science. Like other science writers, he simply kept writing popular books for popular audiences. He had many allies, particularly in Great Britain. Most notably, the Halls worked with New American Library to produce *A Brief History of Science* (1964). This book continued the Halls' quest to celebrate the historical "attack on tradition" as well as the triumph of science over magic and "the elements of the irrational." Marie Boas Hall likewise argued in *The Scientific Renaissance* that "out of the muddled mysticism of sixteenth-century thought and practice, the scientifically valid problems were gradually sifted out to leave only the dry chaff of superstition." A slightly aging generation of historians praised the book widely.[43]

Yet the scene was increasingly dominated by more critical voices. For example, astronomer Fred Hoyle published *Of Men and Galaxies* in 1964. Hoyle lambasted the rise of "big science" as well as the dangers of a powerful scientific elite. The book passionately pleaded for the preservation of romantic settings, where exploration and aesthetic enrichment happened simultaneously. Hoyle argued: "Walk into a big cathedral, and it wipes your brain clean of every thought. . . . The same thing happens all too easily in big science." Big science was stripping away the wonder, enchantment, and awe. These sentiments led the Smithsonian's Walter Cannon to argue that Hoyle "has really gone off the shallow end."[44]

Other writers were taking a critical turn. Most famously, the bestselling works of Rachel Carson pleaded for environmental protection and revulsion at modernist fantasies of the violent redesign of nature. Interestingly, Carson implicitly or explicitly contrasted a romantic appreciation of nature with exploitation, cold logic, and capitalism. As many other intellectuals turned critical eyes toward modernization theories, Western values, and the uncritical celebrations of science, other science writers, like Ley, labored forward or fought against the tide.

More *Rockets, Missiles, and Men in Space*

Ley continued to write about rockets and space travel for adult and juvenile audiences. For example, he edited a collection of governmental reports into a far more readable book, *Harnessing Space* (1963). Each chapter focused on the peaceful uses of artificial satellites in meteorology, astronomy, and communications. The book made the case for the economic and scientific advantages of peaceful space exploration. Many of Ley's other short books expanded on the theme of peaceful exploration.[45]

A more substantial contribution to spaceflight popularization can be seen in Ley's ongoing collaboration with Bonestell, which produced *Beyond the Solar System* (1964). Once again Bonestell received top billing for his artwork. Von Braun also wrote the foreword. As with other collaborations, Ley supplied the text. *Beyond the Solar System* is by far the most speculative of Ley's collaborations with Bonestell. The first chapter narrated a manned trip to Alpha Centauri. Ley admitted that this trip was not feasible in the immediate future. He presented readers with calculations for a rocket that traveled 10 miles per second. Even at this speed, it would take 68,500 years to get to the nearest solar system. Ley argued, "Since anything we could possibly do within a decade after the moon landing would be hopelessly inadequate, let us at least see what would be needed, even if we don't know how to build it."[46]

For the remainder of the book Ley struggled to find middle ground between scientific reasoning and pure fantasy. His tone reads as hesitant and reluctant. The book also presented an imagined, inhabited planet in another solar system. Bonestell provided a fantastic illustration of this alien world. In other images Bonestell painted alien landscapes, which were almost purely works of science fiction. *Beyond the*

Solar System was the least successful collaboration. Unlike *Conquest* and *Exploration*, it was not formatted in coffee-table size. For the most part, the paintings by Bonestell are somewhat lifeless and dull, compared to earlier works. Reviewers largely ignored the book.

Ley wrote several other short books, such as *Our Work in Space* (1964). Additionally, he produced several books for New American Library. These small pulps included *Missiles, Moonprobes, and Megaparsecs* (1964), *Ranger to the Moon* (1965), and *Mariner IV to Mars* (1966). The pulps updated readers on the latest Space Age developments. Ley voiced much support for NASA, while he explained the goals and purposes of specific missions. He also glorified astronauts as heroes. On the one hand, he continued to voice optimism about a future of American "firsts." On the other hand, his books and articles warned readers about the dangers of Soviet space superiority. He even predicted an imminent "space war." Ley wavered between an anti-totalitarian anxiety about technological gaps and lingering hopes for scientific internationalism.[47]

By far his most influential contribution to the scene can be read in his last revision and expansion of *Rockets*, now retitled *Rockets, Missiles, and Men in Space* (1968). The book contained almost 50 percent new material, which updated readers on contemporary developments. Ley also rededicated the book. Instead of continuing to dedicate it to his wife, Olga, he wrote: "The new version of my book is dedicated to the space explorers of the next generation who will want to know what their fathers thought and did." In the book's foreword, Ley reflected on the evolution of *Rockets*, as it had grown from 288 to 576 pages. Whereas his 1944 edition was fairly evenly split between historical accounts and future predictions, his 1968 edition was entirely historical, with few immediate predictions. Ley wrote, "What you are holding now is virtually all history; the amount of prediction that remains is negligible." He added, "Some people may feel that this is a sad state of affairs—it was so nice to dream. But there is no reason for regret. When all the current projects have been carried out they will form a firm basis on which to build still another set of dreams."[48]

Apart from significant revisions, Ley added a large chapter on "Man in Space." As one might expect, the first American astronauts are described as incredibly brave explorers, who endured stress tests and other physical discomforts. Ley's description of future astronauts was somewhat similar to his earlier predictions in *Space Pilots* (1957). Ley then described John Glenn's 1962 orbital flight in dramatic and heroic terms. Ley delighted in combining a description of the astronaut as a bold adventurer who longed for a breathtaking view of the heavens with a more down-to-earth representation of an engineer. For example, Glenn "really wanted . . . a capsule that was all glass," so that he could marvel at the wonders of nature. Yet after reentry and splashdown, his words "were not the kind later put into the mouth of a hero in a play." Glenn simply remarked, "It was hot in there." Other Mercury astronauts used similar language. In Ley's account, American astronauts have interesting personalities. Russian cosmonauts are simply names, without personalities.[49]

The most emotional description of this new breed of explorers occurs in "Postscript: 'If we die . . . '" Ley begins by quoting astronaut Gus Grissom, who told the press: "If we die, we want people to accept it. We are in a risky business. . . . The conquest of space is worth the risk." Ley stated, "The tragic fact is that Grissom did die, along with Edward H. White II, who had been the first American to leave a spacecraft in flight, and Roger B. Chaffee, who was still looking forward to his first trip above the atmosphere." Ley then described the fatal circumstances of their death during a simulation, when a spark of electricity ignited the cabin. According to daughter Sandra Ley, the only time she saw her father cry was during television reports of the disaster. Implicit in Ley's narrative is a deep respect for the daring explorers who risked their lives for the conquest of space. Ley ended his new version of *Rockets* with a celebration of human exploration: "Of course there is no proper ending to the story of rockets and spacecraft to come. . . . The exploration of space will go on forever and ever." The astronauts would be the new heirs to a long tradition of scientific exploration. Their voyages of discovery were beginning. Their frontier was endless.[50]

Although historians of technology ignored the book, the reception of Ley's final revision of his bestseller was mostly positive. One reviewer labeled the book as "his history-cum-encyclopedia." They generally agreed that Ley's new edition was clear and definitive. Reviewer

George Earley noted: "If you can buy only one book to learn not only the past but the future of space flight, this is the one to buy—accept no substitutes." NASA's first historian, Eugene M. Emme, noted, "Having gone through 21 printings and 4 complete revisions since it first appeared . . . Ley's now standard history has a history of its own." According to a review in *Library Journal*, Ley's new edition was "truly monumental . . . the complete record of rocketry and its place in today's world." A reviewer for the *Houston Post* argued, "Every field has a best reference book for laymen, and this is the one for those interested in the history of space flight."[51]

Ley continued to occupy a privileged role as a prominent expert and popularizer of spaceflight. Nevertheless, he had a broader agenda.

Dawn of Zoology (1968)

Throughout much of 1967 and 1968 Ley wrote a book that would serve as his second opus on the history of science. He titled it *Dawn of Zoology* (1968), because it explored the prehistory and evolution of human thoughts about the natural world and its curious creatures. Editor and science writer-naturalist Joseph Wood Krutch introduced Ley's contribution to the Prentice Hall series on Nature and Natural History. Krutch made no apologies for including a "rocket expert." He advised readers to forget what they may or may not know about Ley. Instead, they should consider him "only as a student of the development of zoology." Krutch also outlined Ley's clear agenda in countering a recent and trendy thesis in the history of science: "Mr. Ley avoids the too familiar thesis that all the sciences grew out of the pseudo-sciences— chemistry out of alchemy; astronomy out of astrology; zoology out of myth, fable and the search for moral meaning in natural phenomenon." Instead, Ley would show that "the desire to satisfy curiosity which had no ulterior purpose is the real father of zoology." In other words, Ley presented a no-holds-barred attack on the "nonsense" that "real" science owed a debt to the irrational mysticism of medieval practices or the lingering stumbling blocks, blind alleys, and wrong turns. Ley would chart the gradual ascent of knowledge about the natural world as well as the general transformation of "man the hunter" into "man the explainer."[52]

Ley begins by taking readers into the furthest reaches of historical time in northern Europe, when the "glaciers of the last stage of the Ice Age had then only recently withdrawn." Ley added, "A casual observer from another planet would have seen quickly that Man was on the road to mastery . . . due to the greater size, ability, and efficiency of the human brain." While he did not discount the contributions of hunting and breeding amid a growing awareness of the animal kingdom, he dismissed such practices as utilitarian. He argued: "The prehistory of zoology began with something entirely different—the discovery that there were curiosities." People began to marvel. They also began to explore. Much later, with the Greeks, science began "with wondering."[53]

Unfortunately, the broader quest to explore the world started with Greek historian and writer Herodotus, who "did not do much investigating of his own." Although he was a "careful reporter," Herodotus "did not think it necessary to investigate stories about animals in detail; he may have felt that while somebody might lie when it came to politics or to finance, nobody would say something untrue about such a relatively unimportant thing as animals of a region." In Ley's mind Herodotus did more harm than good. He argued, "Herodotus is responsible for quite a number of ridiculous stories, most of which were firmly believed for centuries to come."[54]

Fortunately the scene would change when Aristotle "studied animals for the sake of knowledge." Ley praised Aristotle's independent thinking, after he broke with Plato and "began to pay more and more attention to facts." Aristotle "collected, dissected, and studied." Other brave thinkers followed. Notably, Pliny the Elder carried the prehistory of zoology into a new millennium. He exemplified "man the collector," who "took stock of what was known." His *Historia naturalis* represented his tireless efforts to write an encyclopedia about the natural world. In spite of its inaccuracies, as well as the uncritical inclusion of mythological creatures, Pliny the Elder's text served as an "indispensable book." It documented the efforts of a naturalist "on the trail of the unusual and remarkable." He offered his readers "a plethora of fabulous beasts."[55]

Ley then discussed the transition to "man the allegorizer," whose fictional fables, poetry, and other stories did little to aid the science of zoology. For example, the "sermonizing of the Physiologus must have struck the right note . . . [but as] far as the 'facts' are concerned it is without value and it certainly is not a literary or poetic masterpiece." During the subsequent medieval period, darkness fell. Ley does not miss an opportunity to write about the tragedy that hindered zoology. The medieval scholars were blinded to the world around them. They followed dogma uncritically. They possessed no real sense of wonder or sheer curiosity. Science could not progress. Instead, the age of darkness elevated "man the cleric."[56]

Despite such sentiments, Ley offered some nuance. He clarified that the term *Dark Ages* obscured the complexities of the era. He wrote, "Everybody 'knows' that no advances of any kind were made . . . that learning was held to be without value and was discouraged and even actively suppressed." Ley added, "Everybody knows that a man with a new idea, like Christopher Columbus, had to fight superstitions and wrong geographical beliefs." Ley then tried to add finer points to the discussion. In a surprising turn, he noted how many key inventions occurred during the era. He also highlighted a "whole catalogue of glittering names," such as Albertus Magnus, Roger Bacon, St. Thomas Aquinas, and Dante Alighieri. Nevertheless, darkness reigned.[57]

Additionally, Ley quoted historian Sarton's description of Hildegardis de Pinguia as a nun who possessed "an encyclopedic mind of the mystical type." Ley took an in-depth look at St. Hildegard's *Physica* as a medical text. He obviously enjoyed poking fun at the work's surprising prescriptions. Ley then examined her discussion of the animal kingdom. More than many writers, she contributed to myths about the unicorn. Ley also implied that she invented new details surrounding the unicorn's fascination with virgins, particularly with young and attractive ladies of the nobility. Overall, Ley dismissed her contributions in a chauvinistic tone: "She was, no doubt, a 'holy abbess' with the common people and the minor clergy, as well as the nobility of her time. But the world of learned letters paid no attention to her."[58]

Ley's description of Emperor Frederick II was kinder, yet he argued that Frederick was "probably the only man who was excommunicated three times. . . . As a person he was generous and expansive—and was promptly accused of being given to orgies." However, Frederick was also an avid learner who possessed "an absolutely insatiable curiosity, embracing everything from astronomy to zoology, especially zoology." Ley described a typical day in Frederick's life, which included "checking on edicts, correcting a translation from Arabic made by one of his scholars, dissecting a bird, and dictating letters to Moslem rulers." Ley presented Frederick as an internationalist in search of universal truths. Ley wrote, "His extensive correspondence with non-Christian rulers made him suspect, his unusual experiments even more so, and his insistence on a weekly bath was probably due to un-Christian influences too."[59]

The remainder of *Dawn of Zoology* celebrated the triumph of human curiosity during the Renaissance and beyond. As one would expect, the book moves from founding father to founding father, while glorifying wonder, mystery, awe, and the liberation of the human spirit from dogmatic systems of belief. Ley's history of zoology is thus a history of the ascent of humankind. He explored many of the key primary sources in order to advance his central thesis: modern science emerged when human beings cast aside ulterior motives and authoritarian cults. The scientist embraced pure curiosity for curiosity's sake. This distinction, in the end, was Ley's counterattack to recent "trendy" theses that blurred the lines between science and pseudoscience. The mystics had suspect motives and authoritarian masters. The scientists had pure motives and few restraints. Thus the history of science offered a lesson in hope, liberation, freedom of thought, and the international exchange of ideas beyond borders and "curtains." Real science was heroic.

The Sophistication of an Academic Discipline

Ley's *Dawn of Zoology* was favorably received by scientists and generally ignored by historians of science. For example, a high school biology teacher praised the book as a useful text for assigned reading, while *Isis* did not solicit a review. When other commentators reviewed the book, their tone was incredibly patronizing, indicating a complete dismissal of the role of a science writer as a historian of science. Most notably, Jerry Stannard reviewed the book for *Science*. He called the book "a popularization, suitable as a Christmas gift but valueless to the scholar." Historian Theodore M. Brown also dismissed the book as a curious artifact of the past. Brown argued, "He seems to write out of the anachronistic conviction that science is mainly the accumulation of factual knowledge and that the history of science, as a result, ought to be devoted to the recounting of past 'errors' and first 'true discoveries.'" Brown continued: "Following this generally (and rightfully) abandoned conception of the task of the historian, Ley develops a chatty, semi-encyclopedic format." Apparently Ley showed little awareness of the history of science as a "socially structured" and "historically evolving activity." It is difficult to find other reviews of the book. Prentice Hall's less than stellar marketing campaign did not help sales. Yet, considering that more than a thousand university libraries possess the book today, it must have circulated quite widely. It was also translated into French. Nevertheless, *Dawn of Zoology* was a personal failure for Ley. It was one of his most direct contributions to the history of science, and historians ignored it.[60]

Not only was Ley's style of writing outdated, but his historical celebration of founding fathers was at odds with the ideals and standards of an emerging and academic profession. As the discipline of the history of science developed, it remained open to academics of diverse training in both scientific and humanistic disciplines. Yet it became less open to both science writers and popularizers. Contemporary historians have celebrated this change for many good reasons. With the professionalization of the history of science, scholars began to tell more complex stories that chipped away at the caricatures of grand narratives and Enlightenment tropes. Medievalists, in particular, followed in the footsteps of Pierre Duhem (1861–1916) by discrediting the notion that the period was devoid of intellectual sophistication. Other scholars spent decades challenging the "classical view of science," along with its traditional dichotomies between science and pseudoscience. There was little enthusiasm for uncritical and celebratory accounts, particularly when embellished histories circulated widely outside the bounds of peer review. There was overt hostility toward practicing scientists who study history. All-encompassing visions of progress reeked of naïveté, or they simply represented ideological justifications for the social, political, and economic importance of science and its practitioners.[61]

This divergence between scholarly historians and science writers would increase in the coming decades, although a few influential figures would continue to navigate a middle world. Yet by the 1970s the contrasts could be jarring. In a retrospective article about the science writers, writer Keay Davidson reflected on the sudden dichotomy. On one side stood Thomas Kuhn, whose works were popular, yet intensely critical of Whiggish agendas and uncritical celebrations of science. Kuhn's many admirers and fellow travelers sought to purge their field of pseudo-history. On the other side stood Carl Sagan, who "viewed science history as a saga of heroes vs. intellectual bigots and progress vs. superstition. . . . One might describe his historical writings as a mixture of Sarton and Andrew Dixon White, plus a pinch of Arthur Koestler."[62]

As an academic discipline, the history of science became far more self-aware and eager to celebrate its liberation from an earlier generation's baggage. By the 1980s historian Jan Golinski summarized: "It was no longer possible to evade the conclusion that the traditional understanding of science had been radically undermined." Much of this progress should be attributed to the scholars of the 1960s. They challenged traditional narratives, cast a critical eye, and asked demanding questions that upset traditional hierarchies of knowledge. A counterculture revolted.[63]

Although it is easy for historians to celebrate these developments, perhaps the example of Willy Ley might lead to less congratulatory accounts of the institutionalization of an academic field. Certainly it is easy and perhaps healthy to cringe when reading his histories of science today. Yet it is also possible to pause and appreciate his works and his broader agenda. Sympathetic historians could ask a provocative question: What was lost in the process of professionalization, when the history of science transitioned from an open and cosmopolitan scene to an academic and institutional setting?

Other scholars have asked this question about their fields. For example, George A. Reisch's *How the Cold War Transformed the Philosophy of Science* (2005) took a fresh look at the early empiricists. Reisch discovered that many positivists sought "to cultivate epistemological and scientific sophistication among even ordinary citizens." Rather than promoting a detached objectivity and a value-free science, they campaigned for hearts and minds, while promoting value-laden justifications for scientific thinking. Reisch further illustrated how later critics deeply misunderstood or mischaracterized these activities and beliefs.[64]

Thus we can see a transition in the philosophy of science. During the 1930s and early 1940s, the philosophy of science flourished as a cosmopolitan and open discourse. Intellectuals became socially engaged, often writing for popular audiences. Yet upon philosophy of science becoming a more respectable academic discipline during the Cold War, this commitment to social outreach declined, due to both academic isolation and the fact that populist activities of leftist intellectuals reeked of a past association with the Popular Front. The rise and fall of public intellectuals marked the emergence of their discipline, especially when a redefined logical empiricism of the 1950s tried to purge itself of social engagement and politics. In the transition to value-neutrality, the "cultural and social ambitions were lost."[65]

Is it possible that this perspective may also apply to the history of science? While most historians of science no longer see the world in stark terms, perhaps the time has come to seek guidance from an earlier generation of writers who had a social cause to construct a world that was hospitable to science, democracy, progress, and reason. Despite their Whiggish interpretations, ahistorical outlooks, and general misconceptions about the history of science, they held very clear notions about the friends and foes of truth. They chose their battles wisely. They celebrated the incredible accomplishments of the past and present. They inspired hope for the future. Like Willy Ley, they influenced millions of people to hope, dream, and act. They had a very real impact on the world in which they lived.

Epilogue

First Citizen on the Moon

Ley died of a massive heart attack on June 24, 1969, weeks before the lunar landing of *Apollo 11*. He was sixty-two years old.

For months he had been working long hours as he struggled to fulfill several outstanding book contracts, including a work on the scientific and literary history of the moon. This project originally sought to capitalize on the lunar landing in July. Chapters included "The 'Astronomical Revolution'" and "Progress: Step by Step." Ley had fallen behind schedule. He still needed to write other chapters, such as "Twentieth Century Moon Madness" and "The Moon from Space." Additionally, he had taken out too many book advances with competing publishers. He was overworked, as other royalties diminished. By this point in his life he had gained much weight, while he continued to smoke cigars.[1]

Despite these warning signs, his family, friends, and colleagues were shocked by his passing. Consequently, their celebrations of the conquest of the moon were bittersweet. His wife, Olga, stated that the moon landing "was the justification of all his dreams." Yet he did not live to witness the live television coverage. For Isaac Asimov, the situation was tragic and ironic: "Willy had spent almost his whole life wrapped in rocketry. He was the world's leading writer on the subject and from his teens he had had the one overriding ambition to see human beings on the moon—and he died six weeks before the attempt was to be made." Science writer Walter Sullivan reported: "Willy Ley, who helped usher in the age of rocketry and then became perhaps its chief popularizer, died yesterday. . . . Mr. Ley lived to within one month of the scheduled fulfillment of his prophecy." Fritz Lang wrote a more emotional obituary. Lang praised Ley's efforts to make a dream become a reality. The lunar landing, Lang implied, was the realization of Ley's dreams: "For him as to me, this day is the symbol of hope—the hope that other dreams born in the minds of other men, good dreams for a better future—will eventually become reality."[2]

Spaceflight magazine presented several tributes. A short note from Chesley Bonestell stated bluntly, "Sir,—Willy Ley has gone. He probably did more than anyone else to make the public space conscious and to help man reach the Moon." P. E. Cleator wrote, "Ley, in one way or another, was so much a part of the astronautical scene that it is difficult to visualize it without him . . . his name will live on as one of the pioneers of that once supposedly fantastic enterprise, the making of a journey to another planet." A. V. Cleaver even foresaw dire consequences for NASA: "With the death of Willy Ley . . . one of the greatest interpreters of astronautics was lost to mankind, at a time when the need for better understanding of the aims and possibilities of spaceflight is critical." Although Ley was on the outside of the institution, his loss would have a dramatic effect, unless a new popularizer could take his place.[3]

Science fiction writer Lester del Rey wrote the most emotional obituary on the night before Ley's funeral. He lamented, "The world of science fiction has just lost its most important citizen. And—if histories are written by men of understanding—it may some day [sic] be realized that the world has lost one of its singularly great leaders, surely a fact not readily apparent to some during his life." Del Rey further praised Ley's lasting contribution to the field of science fiction: "His popular books and articles on the hard facts of rockets, orbits, and space travel established the basic handling of such subjects. Writers who never read any of his early books on rocketry derived most of their facts from him through the stories by men who had studied his writings." Ley's legacy went beyond his influence on science fiction: "He took what must be the very basic dream with which science fiction began. . . . And more than any other man, often by the least obvious means, he built that dream into reality."[4]

Del Rey added, "He was more than a prophet without honor. Events did not pass him by. Rather, he shaped them." According to this perspective, Ley had engineered the Space Age: "It was largely Willy's work that killed the public antipathy to rockets . . . and began to make people dream of space again." More than any other figure, Ley was a visionary who influenced public opinion. "Somehow, through all his articles," del Rey concluded, "Willy and those who were converted by him had managed to convince half of the nation that there was value enough in the space program for them to go along with the huge expenditure. . . . And step by step he led them to turn their eyes from this single planet to the vast reaches of space." Del Rey ended his emotional obituary with these words: "It took him forty years and he missed his goal of seeing the first man on the moon by a month. But there is precedent for that. . . . *And Moses went up from the plains unto the Mountain . . . And the Lord showed him all the land . . . And the Lord said unto him . . . I have caused thee to see it with thine eyes, but thou shalt not go over thither."*[5]

Putting aside this hyperbolic, biblical language, one could argue that in some ways, not only did Ley witness the moon landing, but also he reached the moon. In 1928 he had walked on Lang's movie set. In 1969 a commission agreed to name a crater in Ley's honor. In a fitting tribute Ley's lunar crater can be viewed as an intermediary site between craters named after scientists, engineers, and science fiction authors. Del Rey called him "the first citizen on the Moon," due to his status as an outsider, compared to other legendary names on the lunar landscape.

Ley would have been incredibly humbled by the recognition. Throughout his life, he recognized his own importance and influence, while expressing a sense of humility. For example, in January 1968 Ley attended a cocktail party to celebrate the tenth anniversary of *Explorer 1*. He recalled the event: "Everybody that had anything to do with the project—including some who, like me, had only contributed moral support—was present, and the room reverberated with reminiscences." Ley's statement illustrates his modesty in the company of engineers and scientists. One might also recall his 1955 advice for a future biographer: "I . . . would like to be held responsible to some extent for the coming age of space travel." From his early work as a consultant to *Frau im Mond* to his later publications, interviews, and even toys, Ley spoke incidentally about his efforts and successes. He appreciated recognition and some degree of fame. He adored and preserved many of those recognitions, particularly letters from children. Yet he summarized his role in the Space Age as merely contributing "moral support," as the real engineers and scientists did the hard work of overcoming engineering obstacles and designing the marvelous technologies of Tomorrowland. At times he felt left out of the really exciting and creative work. Yet, as he told journalist Mike Wallace in 1957, he had a "good substitute."[6]

By 1969 Ley had written dozens of books that excited audiences about the future of human spaceflight. He had written hundreds of articles for newspapers, magazines, and newsletters. He had given hundreds of public lectures, meant to educate, entertain, and excite a crowd of curious onlookers. His influence could be seen everywhere, from television broadcasts to science fiction novels. Ley influenced every medium in some capacity. In addition to print, film, and broadcast media, his publicity efforts included postcards, toys, exhibitions, rocket rides, and cereal boxes. Von Braun once acknowledged Ley's importance: "During the past thirty years he has done more than any man I know to carry the space message to the public, particularly to the younger generation. He deserves much credit for the space consciousness which has gripped the United States and which is the indispensable foundation of the American space program." Although Ley rarely designed or experimented, he engineered public support for a future of interplanetary travel. In spite of his outsider status, he effectively retained the title of a "scientist" and "rocket expert," who could educate millions of Americans about the field of rocketry and the cutting edge of space exploration. He directly and indirectly shaped both European and American "astroculture," with its blurring array of images, artifacts, and spectacles. He should be remembered as one of the chief architects of the Space Age as well as the movement's chief publicist.[7]

For forty-three years Ley acted as an intermediary who took his readers and audiences into space. Whether as passengers on a Disneyland rocket ride or as readers of *The Conquest of Space*, millions of American children and adults experienced the journey to the moon in imaginative and thrilling ways. Watching the event occur live in 1969 may have been a far more passive experience, especially for young adults coming of age during a time of social unrest, civil rights struggles, and the Vietnam War. The actual moon landing may have seemed anachronistic or out of place. An interesting social study of this group and somewhat older individuals might ask a relevant question: During the moment of touchdown, did you feel immense hope for the future or a lingering nostalgia for a childhood of rocket dreams, space cadets, and 1950s optimism?[8]

Ultimately the Space Age failed to provide experiential fulfillment in a way that satisfied a need for transcendence. Nevertheless, it temporarily satisfied romantic and spiritual yearnings. As scholar Marina Benjamin noted, subsequent decades included a wild combination of cosmic religiosity, mysticism, fanaticism, scientific worship, and technological fetishism. When reality crushed cultural and spiritual ambitions, dreamers began to inhabit cyber or virtual realms that offered fewer constraints, along with more opportunities for cultural rebirth. Unless a miracle technology of the future allows for safe and convenient journeys to other worlds, human beings possess a satisfying alternative. Rocket technologies and even the international space station have been dwarfed by media technologies delivering fantastic results that match social expectations. Arguably, Benjamin's account overprivileges escape, community building, and utopian hopes rather than the fundamental motivations. For Ley, there was a "basic drive" that united all freethinking human beings throughout time and space. The longing for discovery and understanding was the engine of all history, from Babylon to the Space Age. The quest for other worlds and knowledge unified humankind.

This glorification of science and exploration was incredibly modernist and futuristic. In fact, scholars may be tempted to use the term of *astrofuturism*, as coined by scholar De Witt Douglas Kilgore. Kilgore placed Ley firmly in a first generation of astrofuturists, who promoted a common vision of space exploration, ripe with frontier imagery, capitalistic fantasies, and dreams of cultural rebirth. Kilgore claimed that Ley presented "the space frontier as a natural extension of Western and, therefore, American culture." Individuals like Ley either explicitly or implicitly supported a manifest destiny to conquer and subjugate new frontiers. Ley made few distinctions between Western knowledge and universal truth. Accordingly, he promoted an imperialistic vision of conquest.[9]

We can only speculate as to Ley's response to such a critique. Ley would be baffled by derogatory references to progress, which are so fashionable in higher education today. He would scoff at people who associate remarkable human accomplishments with materialistic, imperialistic, or capitalistic quests to conquer other worlds. He would admit that economics, national prestige, and international competition often facilitate exploration. Nevertheless, if states and empires did not exist, a fundamental drive for knowledge and discovery would still exist. The quest for other worlds drives human history. The quest for knowledge fuels progress. Ley would have agreed with the narration of a recent NASA publicity video:

> We are the explorers. We have need to find what is out there. It is a drive inside each and every one of us. The drive to wonder, to push the boundaries, and to explore. . . . New vessels will carry us, and new destinations await us. Everything that we have ever accomplished leads to this moment in time, where exploration will now take us to the planets and the stars.[10]

The video concludes: "We are the explorers. Throughout our history, we have taken both small steps and giant leaps in that pursuit. Our next destination awaits. We don't know what new discoveries lie ahead, but this is the very reason we must go." Our curiosity compels us.

Ley articulated this rationale quite often. For example, during the 1957 *Night-Beat* interview, Mike Wallace asked: "Why do you want to go into outer space? What's your fascination with it?" "Well," Ley answered, "you have the old answer to the question of why do we want to climb Mt. Everest: Because it is there . . . it is a basic drive." Ley added,

"Man was born a curious animal." Wallace then quoted a scholar who related space travel to "symbolic satisfaction for erotic or aggressive needs. It's just as basic as sex . . . the urge to explore, and to manage what we explore is a human urge as fundamental as the urge to procreate." Wallace asked Ley, "Well, what about it?" Ley briefly hesitated, most likely due to the sexual language. Then he responded: "Well he is probably right. I mean, I wouldn't have phrased it this way . . . I probably wouldn't have thought of it this way." Ley then summarized his own perspective by invoking historical parallels: "People of curiosity went after things with amazing results."[11]

Although scholars will continue to analyze frontier imagery, teleological narratives of progress, and utopian hopes, it is important to recognize the matter-of-fact nature of this claim. Ley never spoke of a grand vision for cultural rebirth. Rather, Ley presented a rationale for spaceflight that assumed the form of a syllogism, as described by Stephen J. Pyne: "The urge to explore is a fundamental human trait. Space travel is exploration. Therefore, sending people into space is a fundamental characteristic of our species—what more is there to say?" Von Braun made similar arguments, such as: "Man's restless mind is not easily discouraged by obstacles . . . all obstacles, no matter how high, will ultimately be overcome." According to von Braun, it did not matter that the closest star system was over four light years away. What had seemed fantastic in the 1920s became a reality in the 1960s. Therefore what seemed fantastic and impossible in the 1960s would actually happen in the 2000s. Von Braun added, "A thousand-mile journey begins with a single step." Ley also believed that a manned journey to Alpha Centauri would happen around 2014. It did not matter that something was currently impossible from a scientific or technological point of view. Ley and von Braun expressed faith in the human spirit. They celebrated the future accomplishments of humanity.[12]

Ley, in particular, voiced faith in science as the guiding force of universal progress. Like other "scientistic" thinkers, he would have argued that there were no limits that science could not overcome. He would have agreed that scientific knowledge is the best and most valid type of truth. Secular, rational, and scientific values triumphed over irrational and subjective views. Yet this form of scientism did not necessarily fit well with big science. Indeed, there are deep tensions between a dogmatic materialism that privileges laboratory methodology and Ley's "romantic naturalism." Rarely did Ley speak about a scientific method. He spoke instead about experiential moments of human creativity and curiosity, in which many of the key explorers were rule breakers and system destroyers. Ley celebrated the journey, whether it involved the nighttime explorations of a daring astronomer or the jungle adventures of a brave naturalist. Ley did not reduce all knowledge to individual and measurable parts. He built up from a complex and interdisciplinary web of truths to create a mosaic of the universe that inspired wonder, awe, and moments of spiritual transcendence. Ley sought to know the cosmos, as he was a part of it, and it was a part of him.[13]

This spiritual quest was fueled by unequivocally modernist hopes and dreams of a better tomorrow, as humankind continued to transcend the boundaries and evolve as a biological species. It included grand visions of reshaping the earth or other worlds, of somehow harnessing forbidden resources to alleviate social ills, and of conquering nature for the benefit of all humankind. The story of humanity was an adventurer's tale out of the darkness and into the light. At some point, the "saltation" will occur. Human beings will ascend like the brave creatures that boldly crawled out of the ocean. It will be a story of human destiny. Humans will transcend limitations.[14]

Ley's perspectives fit well with the flourishing of "romantic modernism" throughout the twentieth century. One hopes future scholars will confirm whether this label can be applied more broadly to American popular science and natural history beyond the nineteenth century. Perspectives are moving in this direction. Scholars have begun to question traditional lines of demarcation in intellectual and cultural history. For example, scholar Michael Saler reevaluated a type of attitude that blended modernism and romanticism. He has interrogated the associations of modernity with an allegedly cold, rationalist, and "disenchanted" era. Modernity can be described as an equally enchanted world. Saler also identifies a "larger cultural project of the West: that of re-enchanting an allegedly disenchanted world." Wonder, awe, and the sublime did not contradict a scientific worldview. Instead, they could complement and overlap each other. These combinations revived a genre of science writing and popularization that had existed for centuries.[15]

The Space Age provided a remarkable opportunity for neo-Humboldtian worldviews and transcendental hopes. Scholars have long recognized the romantic and transcendental associations in spaceflight media, while new scholars are continuing to break down traditional distinctions. Much more scholarship must be done before we can make definitive conclusions. Nevertheless, I hope that a detailed portrait of the works and career of one of the most influential science writers of the twentieth century lends support to the claims of other scholars. Ley's brand of popular science mixed genres. He combined science with wonder, reason with faith, reductionism with holism, and technology with subliminal awe. He had a broader cultural project of reenchanting an allegedly disenchanted modernity. While he offered a spiritually fulfilling appreciation of the mysteries of nature and the ascent of humankind, he celebrated the human spirit in a way that put human beings in control of their own destiny. His explorers, as well as his readers, were not the pawns of big science or the victims of a technological maelstrom. The explorers bore little resemblance to Dr. Frankenstein or the Invisible Man. They were not consumed by their power and prestige, while single-mindedly pursuing a quest with disregard for the consequences. They were not haunted or plagued by their successes. Instead, they were selfless explorers driven by an inescapable urge for discovery. They fantasized about the ways in which science and progress would create a new future, in which new discoveries could pay for everything, alleviate the need for conflict, and create brave new worlds. Human agency still mattered. The explorer was in control. Human beings would shape human destiny.[16]

Arguably these tropes can help to explain how the Space Age was exemplified by the image of the astronaut. Despite the centrality of big science and complex networks of organizations and contractors, the heroic figures of the endeavor were adventurers, displaying bravery in the face of danger as they crossed into the endless frontier. Scholars have begun to analyze the image of the astronaut in more complex ways. These scholars have studied images of American astronauts in terms of a postwar crisis of masculinity, the reassertion of frontier imagery, and the continuity with traditional images of flying aces. Not only did the astronaut represent a romanticized hero, but the astronaut also embodied American traditions of individualism, heroism, and self-sacrifice.[17]

At the height of big science, with its NASA technicians and governmental bureaucrats, Americans embraced the image of the astronaut as a space cowboy and fearless adventurer, taming a new frontier. As other scholars note, the early image of Mercury astronauts as trailblazers easily transitioned to representations of more routine settlers. These images fit well with other scholars' analyses of the blending of frontier imagery and futuristic technologies. It would be interesting for a future scholar to bring in the image of the eighteenth-century explorer or the nineteenth-century naturalist. All these tropes could be part of a broader revival of American romanticism during the twentieth century. Or they could simply illustrate continuity in terms of popular culture.[18]

Overall, cultural representations of spaceflight combined nostalgia and futurism, along with other motifs that have been staples of natural history and popular astronomy. The longevity of representations should indicate that American (or transnational) popular culture can provide a starting point to chart an indigenous, amateur, outsider, or people's science that flourished in the twentieth century. It remained open, while it mixed genres. It resisted and fought back against hierarchies of knowledge. It appealed to Americans because it reflected American values. Thus it makes sense that when *Sputnik* created demands for large-scale hierarchies of knowledge production, America embraced representations that reenchanted big science. The most modern accomplishment of science and technology had to still be a romantic adventure with a romantic adventurer. Willy Ley would have agreed.

Notes

Introduction

1. Saler, *As If*, 7.

2. Saler, "Modernity and Enchantment," 696. See also Merchant, *Death of Nature*.

3. See Saler, *As If*, 12. See also Nye, *American Technological Sublime*; Gilbert, *Redeeming Culture*.

4. See Von Engelhart, "Romanticism in Germany." See also Halliwell, *Romantic*; Holmes, *Age of Wonder*, 468–69.

5. Sachs, *The Humboldt Current*, "deep feeling" 13, "Nature offered" 45. See Walls, *The Passage to Cosmos*; Stephen J. Pyne, "Seeking Newer Worlds: An Historical Context for Space Exploration," in Dick and Launius, eds., *Critical Issues in the History of Spaceflight*, 23.

6. Sachs, *The Humboldt Current*, 13; see also Pyne, "Seeking Newer Worlds," 25; Richards, *The Romantic Conception of Life*, 200.

7. For classic examples, see *History of Science* 32, no. 3 (1994): 237–67. For more recent and quite ground-breaking works, see James Secord, *Victorian Sensation: The Extraordinary Publication, Reception, and Secret Authorship of Vestiges of the Natural History of Creation* (Chicago: University of Chicago Press, 2000); Aileen Fyfe and Bernard Lightman, eds., *Science in the Marketplace: Nineteenth-Century Sites and Experiences* (Chicago: University of Chicago Press, 2007); Ralph O'Connor, *The Earth on Show: Fossils and the Poetics of Popular Science, 1802–1856* (Cambridge: Cambridge University Press, 2004).

8. This biography lends further support for the claims within John Gatta's *Making Nature Sacred*. See also Lightman, "'The Voices of Nature,'" 187–89;

Lightman, *Victorian Popularizers of Science*; Bowler, *Science for All*. On female naturalists see Shteir, *Cultivating Women, Cultivating Science*; Judd, *The Untilled Garden*.

9. Daum, *Wissenschaftspopularisierung im 19*; see also Kelly, *The Descent of Darwin*; Harrington, *Reenchanted Science*; Nyart, *Modern Nature*; Daum, "Varieties of Popular Science" and "Science, Politics, and Religion."

10. Burnham, *How Superstition Won*. See also the observations of LaFollette in *Science on American Television, Making Science Our Own*, and *Science on the Air*.

11. This perspective will be related directly to the recent works of Andrew Jewett and John L. Rudolph. See Jewett, *Science, Democracy, and the American University*; Rudolph, *Scientists in the Classroom*.

12. Other scholars have explored the democratization of knowledge. See, for example, Gilbert, *Redeeming Culture*; Pandora and Rader, "Science in the Everyday World."

13. On science in the vernacular, see Leane, *Reading Popular Physics*; Topham, "Rethinking the History of Science Popularization."

14. Parrish, *American Curiosities*; Butsch, *The Citizen Audience*. See also Butsch, *The Making of American Audiences*; Starr, *The Creation of the Media*.

15. See Secord, "Knowledge in Transit."

16. On the evaluation of von Braun, see Launius, "The Historical Dimension of Space Exploration," 25. See also Neufeld, *The Rocket and the Reich* and *Von Braun*; Launius, *Frontiers of Space Exploration*, 190. For the best historiographical review of the field of American space history, see Asif S. Siddiqi, "American Space History: Legacies, Questions, and Opportunities for Future Research," in Dick and Launius, eds, *Critical Issues*, 434.

17. On Ley as a founding father, see Neufeld, "Creating a Memory," 71.

18. On tropes of spaceflight history, see Siddiqi, "American Space History," in *Critical Issues*, 436. See also Launius, "The Historical Dimension." For a critique of Ley, see Kilgore, *Astrofuturism*.

19. This perspective builds most directly from Alpers, *Dictators, Democracy, and American Public Culture*. See also Hollinger, "Science as a Weapon in *Kulturkämpfe*"; Hollinger, *Science, Jews, and Secular Culture*.

20. For "faith in science" quotation, see Lundberg, *Can Science Save Us?*, 115.

CHAPTER 1. YOUTHFUL HORIZONS

1. For the detailed account of his classroom experience, see Ley, *Willy Ley's Exotic Zoology*, vii. For figures see Audoin-Rouzeau and Becker, *Understanding the Great War*, 22. See also Vinen, *A History in Fragments*, 70 and 75.

2. See Davis, *Home Fires Burning*, 43, 64–65.

3. For the 1955 autobiographical account, see Anonymous, "Ley, Willy," *Twentieth Century Authors First Supplement*, 580. The tale of Willy's mother is told in

Moskowitz, "The Willy Ley Story," 32. See also Moskowitz, "Willy Ley: Forgotten Prophet," 13.

4. Anonymous, "Ley, Willy," *Twentieth Century Authors First Supplement*, 580. For childhood recollections see, for example, Ley, "For Your Information," October 1969, 101–8, 113, quote at 101.

5. Ley, *Mars, der Kriegsplanet*, 5. See Fischer, *The Empire Strikes Out*, 11. See also Ley, *Rockets and Space Travel*, 49.

6. Ley, *Rockets* (1944), 42. Here Ley is speaking most directly about a novel by Achille Eyraud, before adding, "It was the same feeling which produced Jules Verne."

7. Haynes, *From Faust to Strangelove*, 129–35. See also Ley, *Rockets* (1947), 42.

8. Ley, *Rockets* (1947), 42.

9. Scholarly quotations in this paragraph are taken from Chesneaux, "Jules Verne's Image of the United States," 111–13.

10. Adas, *Machines as the Measure of Men*, crisis of Western Civilization 365, "Little that was glorious" 371; Ernst Jünger, "Materialschlacht," *Standarte* 1, no. 5 (October 4, 1925), quoted in Leed, *No Man's Land*, 154.

11. On Ley's recollections see Ley, "Are We Going to Build a Space Station?" 133. See also Fritzsche, *A Nation of Fliers*, 64; Ernst Jünger, *Werke*, I, 368, quoted and translated in Bullock, *The Violent Eye*, 152; Leed, *No Man's Land*, 136.

12. Ley, "Willy Ley Recalls 'Captain Future' of Germany." Some scholars could describe a reconciliation under the influential label of "reactionary modernism." See Herf, *Reactionary Modernism*.

13. On the "shadow" see Ley, *Wily Ley's Exotic Zoology*, vii. See also Anonymous, "Ley, Willy," *Twentieth Century Authors First Supplement*, 580; Anonymous, "Ley, Willy," *The Scanner*, 3. For comments on the "special hall" see Ley, "Monsters of the Deep," 99; "first love" comment from Ley, "How It All Began," 24. On subsequent quotations see also Ley, *Dragons in Amber*, 69.

14. Ley, "Death of the Sun," *Galaxy*, March 1955, 54–65, at 54. On his admiration of Humboldt, see Ley, "Any Questions? The Shape of Shells to Come," *Galaxy*, April 1962, 77–87, at 84. Lasswitz translated and quoted in Fischer, *Empire Strikes Out*, 63.

15. Bowler, *Science for All*, 79, 81.

16. See Ley, "The Man I Didn't Meet," 134; Zell, *Neue Tierbeobachtungen*; Zell, *Straußenpolitik*; Zell *Moral in der Tierwelt*; Bowler, *Science for All*, 5.

17. An inventory of Ley's personal library reveals that he possessed nearly every book by Wilhelm Bölsche, each in the original edition, which indicates that these books may have been purchased in Germany and later sent to the United States in 1935. Special thanks are due to Anne Coleman of the University of Alabama–Huntsville. She provided an inventory of Ley's library when sold after his death. See Bölsche, *Das Liebesleben in der Natur* (Florenz and Leipzig: Diederichs, 1898); for quotations see Bölsche, *Love-Life in Nature*, v–vi.

18. Bölsche, *Love-Life in Nature*, 5.

19. Ibid., "You are on earth," 419. For other perspectives see Kelly, *The Descent of Darwin*, 145; Hopwood, "Producing a Socialist Popular Science," 119; Daum, *Wissenschaftspopularisierung im 19*, 451.

20. See Ley, *Wily Ley's Exotic Zoology*, vii. On ambitions see Anonymous, "Ley, Willy," *Twentieth Century Authors First Supplement*, 580; on "truth" see Ley, "The End of the Rocket Society, part 1," 67; see also Cornell, "More Information, Please!" 7.

21. Many biographical summaries claim that he attended both the University of Konigsberg and the University of Berlin. See Moskowitz, "Willy Ley: Forgotten Prophet," 14. The stories are doubtful. See also "Ley, Willy," *Twentieth Century Authors First Supplement*, 580; Ley, *Wily Ley's Exotic Zoology*, viii.

Chapter 2. From the Earth to the Moon, via Berlin

1. Cornell, "More Information, Please!" 7. Ley incorrectly remembered this year as 1922.

2. Moskowitz, "The Willy Ley Story," 32; see also Ley, "End of the Rocket Society, part 1," 67; on his job see Cornell, "More Information, Please!" 7. See also Kater, "The Work Student," 73.

3. See Daum, "Science, Politics, and Religion," 115–22.

4. On tensions and reactionary rhetoric, see Harrington, *Reenchanted Science*, xvi–xvii.

5. On Ley's walk, see Ley, "How It All Began," 23. On center of life reference see Kracauer, "Lokomotive über der Friedrichstraße," 194–95. See also Ward, *Weimar Surfaces*.

6. Gay, *Weimar Culture*, 6–8. On Americanization see Saunders, *Hollywood in Berlin*; Fritzsche, *Reading Berlin*, 3.

7. Vinen, *History in Fragments*, 151, 171.

8. Fritzsche, *A Nation of Fliers*, 137.

9. Breit, "Talk with Willy Ley"; Cornell, "More Information, Please!" 7.

10. See Fritzshe, *Reading Berlin*, introduction; Ward, *Weimar Surfaces*, 197; Killen, *Berlin Electropolis*, 8.

11. Ley, "How It All Began," 23. See also Valier, *Der Vorstoss in den Weltenraum*.

12. Oberth, *Die Rakete zu den Planetenräumen*. On the story of walking home, "shock," "the equations!" and "over and over," see Ley, "How It All Began," 23–24. "As far as the general public" is from Ley, *Rockets* (1944), 106. On Oberth see also Barth, *Hermann Oberth: Vater der Raumfahrt: autorisierte Biographie* (Esslingen: Bechtle, 1991), 73–78, as referenced in Neufeld, *Von Braun*, 24.

13. Ley, *Rockets, Missiles, and Space Travel* (1958), 116. Subsequent quotations are from this page, except "simplify Valier's book," taken from Ley, "End of the Rocket Society, part 1," 67.

14. For quotation, see "Ley, Willy," *Twentieth Century Authors First Supplement*, 580.

15. Ley, *Die Fahrt ins Weltall*.

16. Ibid., 68. In the original text the passage reads with spacing emphasis, not capital letters. See also Neufeld, *The Rocket and the Reich*, 10.

17. Ley, *Rockets* (1958), 116.

18. Ley, "End of the Rocket Society, part 1," 67–68.

19. Ley, *Rockets* (1958), 24.

20. See "Fragebogen für Mitglieder," *Reichsverband Deutscher Schriftsteller*, 25 September 1933, Bundesarchiv, Potsdam, Germany, "Ley, Willy," RKK:2100, box 0239, file 008. Translation mine.

21. For an example of Ley's *VB* articles see Willy Ley, "Die große Seeschlange," *Völkischer Beobachter*, inside section: "Politik/Belehrung/unterhaltung." For the Munich rally story see Campbell, *John W. Campbell Letters*, vols. 1 and 2, indexed Ley. On correspondence with Heinlein, see Willy Ley to Robert A. Heinlein, 18 July 1940, "Correspondence, Pre-War, 305J," 2, Robert A. and Virginia Heinlein Archives, University of California, Santa Cruz.

22. Ley, *Mars, der Kriegsplanet*, 3–4. All quotations in this paragraph are from these pages except "understandable description," taken from Advertisement, *Die Rakete*, May 1929, inside cover. The story of the Mars exhibition is told in *Mars*, 5. Translation mine.

23. Ley, *Mars, der Kriegsplanet*, 52–55.

24. On Valier's occult writing see, for example, Valier, *Die Sterne Bahn und Wesen*. For a biography of Max Valier, see Essers, *Max Valier* (1968), also available as a NASA technical translation (1976). All quotations in this paragraph are from Ley, "How It All Began," 25.

25. Ley incorrectly reported that these events occurred in the month of June. See Winter, *Prelude to the Space Age*, 15. "Our sights" from Ley and von Braun, *The Conquest of Space* (recording), 00:23–00:24, accessible at archive.org; "The purpose" from Winter, *Prelude*, 35; "spreading the thought" from Ley, *Rockets* (1944), 113; 1940 comment from Ley, "What's Wrong with Rockets?" 49.

26. On Winkler and Valier's aims, see Ley, "The End of the Rocket Society, part 1," 65; see also Freeman, *How We Got to the Moon*, 34–35.

27. For Ley's comment on the growth of the organization, see Ley, *Rockets* (1958), 118; see also Roger D. Launius, "Prelude to the Space Age," in Logsdon et al., *Exploring the Unknown*, 9.

28. Willy Ley, *Das Drachenbuch: Plaudereien von Echsen, Lurchen und Vorweltsauriern* (Leipzig: Thüringer Verlags–Anstalt H. Bartholomäus, 1927). For quotations, see page 6. On earlier appendix, see Willy Ley, *Eiszeit* (Erfurt: Thüringer Verlags–Anstalt H. Bartholomäus, 1927).

29. For "plan" and claim of Goddard correspondence, see Ley, "The End of the Rocket Society, part 1," 68. See also Ley, *Die Möglichkeit*.

30. For original goals, see preface of *Die Möglichkeit*. The last quotation in this paragraph is from p. 340.

31. Translation taken from Freeman, *How We Got to the Moon*, 42. For original text, see Ley, *Die Möglichkeit*, iii–iv.

32. See Ley, *Rockets* (1958), 119.

33. See Winter, *Prelude*, introduction; Neufeld, "Weimar Culture and Futuristic Technology." See also Neufeld, "Spaceflight Advocacy from Weimar to Disney," 72–76; Neufeld, *Von Braun*, 25; Crouch, *Aiming for the Stars*, 42, 48–50.

34. "Rocket-minded" from Willy Ley, Galley Proof for *Inside the Orbit of the Earth*, Willy Ley Collection NASM Udvar-Hazy, box 5, folder 5, 26. See also Neufeld, *Von Braun*, 30; Neufeld, "Weimar Culture and Futuristic Technology"; Crouch, *Aiming for the Stars*, 50.

35. For the engineer's announcement, see unknown author, "Das Raketenauto rast!" 15. Opel's translation taken from unknown author, "Rocket Auto Tops 2 Miles a Minute," 6.

36. On "big, carefully staged show," see Ley, *Rockets* (1944), 115; "plaything" from Ley and Cleator, "The Rocket Controversy," 19; "Powder rockets" from Ley, "What's Wrong," 39; "We had gone" from Ley, "End of the Rocket Society, part 1" 69; "Small wonder" from Ley, *Rockets* (1944), 118.

37. On Opel's "opportunity" see Ley, "End of the Rocket Society, part 1," 69; "Headline stunts" from Ley, "How It All Began," 49; see also Ley, Galley Proof for *Inside the Orbit of the Earth*, WLC, 27. On the expulsion of Valier from Ley's book, see Ley, "End of the Rocket Society, part 1," 69; on VfR membership rewards, see Winter, *Prelude*, 36–38; Ley recounts the support of Oberth in Ley, *Rockets* (1958), 122.

38. Quotations in this paragraph are from Ley, Galley Proof for *Inside the Orbit of the Earth*, WLC, 28.

39. Ley, *Konrad Gesner*, 308.

40. Ibid., 310. Ley's dismissal of Islamic science was widely shared in the early twentieth century.

41. Neufeld, "Weimar Culture and Futuristic Technology," 727.

42. On "paying lip service," see Forman, "Scientific Internationalism," 152; subsequent quotations from 180, 154. Ley's postcards can be found in the G. Edward Pendray Papers at Princeton's Mudd Manuscript Library, 1930s folders.

43. "I did more and more" from Ley, "End of the Rocket Society, part 1," 68; later interview to be heard in Ley and von Braun, *The Conquest of Space*, 10:52–10:57; see also Ley, *Rockets* (1958), 117.

44. Quotations from unknown author, "Probekapital aus Ley," 60–61, except Oberth's "Vorwort," from Ley, *Die Fahrt ins Weltall* (1929 edition), 3.

45. For quotations see Ley, Galley Proof for *Inside the Orbit of the Earth*, WLC, 29.

46. Ley, *Rockets* (1958), 124.

47. On Max Valier, to be heard in Ley and von Braun, *The Conquest of Space*, 3:35–3:44; for Oberth's turmoil and "mystic inclination" see Ley, *Rockets* (1958), 125; quotations on Oberth's "mental make-up" and disapproval of Berliners are from Ley, "End of the Rocket Society, part 1," 72; "astonished disbelief" from Ley, *Rockets* (1958), 126; "missed appointments" from Ley, *Rockets* (1944), 122; on Oberth's Nazi membership, see Ley, *Rockets* (1944), 122, footnote.

48. Quotations from Ley, "End of the Rocket Society, part 1," 70–73.

49. Quotations from Ley, *Rockets* (1958), 127–30.

50. Ley, *Rockets* (1944), 121–22.

51. Ley's crediting of Oberth is found in many editions of *Rockets*. This quotation is taken from Ley, Galley Proof for *Inside the Orbit of the Earth*, WLC, 29. "Only an author" from Hermann Oberth, "Interview with Martin Harwit and Frank Winter," November 14 and 15, 1987, 33, courtesy of Frank Winter, winterf@si.edu; see also Winter, "Frau im Mond," 40.

52. Lang, "Sci-Fi Film-maker's Debt to Rocket Man Willy Ley."

53. Ibid.

54. Ley, "Eight Days in the Story of Rocketry," 61; Oberth, *Wege zur Raumschiffahrt*.

55. On "demand" see Ley, *Rockets* (1944), 125. "I got involved" from Ley, "How It All Began," 50; "I wrote" from Ley, *Rockets* (1944), 125; see also Ley, "Frau im Mond," 13–16; on kaleidoscopes see Franz Storch, "Hermann Oberth und die Frau im Mond," *Neue Literatur* 20, no. 9 (1929): 52, translated in Freeman, *How We Got to the Moon*, 45; see also Winter, "Frau im Mond," 40.

56. Quoted and translated in Freeman, *How We Got to the Moon*, 46. Original source: Willy Ley, "Berlin spricht vom Raumschriff," *Die Rakete*, November–December 1929, 27–28.

57. Ley, "Berlin spricht," 27–28.

58. On the role of imagination see Stephen J. Pyne, "Seeking Newer Worlds," in Dick and Launius, *Critical Issues in the History of Spaceflight*, 7–36, at 8.

59. For a critique of Oberth and Ley's optimism, see Neufeld, "Weimar Culture and Futuristic Technology," 731.

CHAPTER 3. DEATH OF A SCIENCE IN GERMANY

1. On Ley's exchanges with GIRD, see Ley to Pendray, 27 July 1932, Pendray Papers, box 4, folder 12, 1, Princeton University. For an example of Ley's articles in *Astronautics*, see Ley, "The Why of Liquid Propellants for Rockets."

2. For detailed information about Ley's first science fiction novel, see Ley et al., *Die Starfield Company*. See also Nagl, "National Peculiarities of German SF," and "SF, Occult Sciences, and Nazi Myths"; Fisher, *Fantasy and Politics*, 6.

3. Gail, *Der Schuss ins All*, 30–36.

4. Anton, *Brücken über den Weltenraum*. For translation, see "Interplanetary Bridges," *Wonder Stories Quarterly* 4, no. 2 (1933): 114.

5. For *Die Rakete* quotation see Ley, *Rockets* ((1951), 131–32.

6. See Ley, *Rockets* (1958), 126–27; on Nebel as a "professed militarist," see Ley, *Rockets, Missiles, and Men in Space* (1968), 117; for a counter perspective, see Nebel's autobiography, *Die Narren von Tegel*. See also Ley, *Rockets* (1968), 121. For quotation on work, see Ley, "End of the Rocket Society, part 1," 70.

7. Ley, *Rockets* (1958), 135–36, footnote.

8. On test firing and launch figures, see Winter, *Prelude*, 39–43. Einstein's son-in-law's reaction is quoted on page 42, attributed to Marianoff and Wayne, 1944, 115.

9. On program and "embryonic spaceship," see Ley, *Rockets* (1958), 139; "no nonsense," Ley, *Rockets* (1944), 136; "honest and very serious," Ley, "What's Wrong," 39.

10. See Ley, "The Reader Speaks."

11. Ley, "End of the Rocket Society, part 2, 69–70. All quotations are found on these pages except Ley's on a "psychological explosion," found in Ley, "End of the Rocket Society, part 1," 53.

12. On demonstrations see Winter, *Prelude*, 40. For "real progress" and "for publicity" quotations, see Ley, Galley Proof for *Inside the Orbit of the Earth*, WLC, 30, 43. For other quotations see Ley, "End of the Rocket Society, part 2," 62, 68, 69.

13. On Pendray's visit, see Pendray Papers (hereafter cited as PP), box 3, folder 16, "Early, 1931," 1. See also Ley, *Rockets* (1944), 137, footnote. For more on Pendray, see Cheng, *Astounding Wonder*. "Bluff only" from Willy Ley to G. Edward Pendray, 5 January 1932, PP, box 4, folder 12, 2. On shared technical information, see Ley to Pendray, 25 May 1931, PP, box 3, folder 16, "Early 1931" 1. On postcards see, for example, Ley to Pendray, 4 December 1932, PP, "Ley Willy, 1932."

14. Ley's acknowledgment of reading Goddard and other statements can be found in Willy Ley to G. Edward Pendray, 20 August 1931, PP, "Early Rocketeers—Willy Ley, Biography and Correspondence, 1931," 1–3.

15. For quotations on Oberth, see Ley to Pendray, 4 May 1931, PP, "Early Rocketeers," 1–2; on Lasser's *Conquest*, see Ley to Pendray, 10 October 1932, PP, box 4, folder 12, 1. *Starfield Company* correspondence can be found in Ley to Pendray, 23 July 1931, PP, "Early Rocketeers," 1.

16. Quotation regarding the police from Ley, *Rockets* (1958), 151; for "Vorstand" passage, see manuscript, Willy Ley, "Around European Rocketry," PP, box 4, folder 12, 1. In a later issue of *Startling Stories*, Ley referenced "The 'Magdeburg Rocket' of the German Rocket Society's research group, finished in June 1933." See Ley, "The Road to Space Travel, Part 1: The Last Twelve Years," 71.

17. Ley, *Rockets* (1958), 152.

18. On "deterioration," see Ley, *Rockets* (1958), 153–54; other quotations from

Ley, *Rockets* (1944), 150. See also Ley quoted in Sharpe and Ordway, *The Rocket Team*, 16.

19. On blitzkrieg of publicity, see Ley, "End of the Rocket Society, part 2," 59. See also Ley, *Rockets* (1958), 152–53; "Around the middle" from Ley, "End of the Rocket Society, part 2," 59. For Pendray correspondence, see Ley to Pendray, 5 January 1932, PP, box 4, folder 12, 1.

20. On "pure comedy, " see Ley, *Rockets* (1958), 155. This page also contains the "inadequate and senseless" passage. For the description of the "busily plotting German army," see Ley, "End of the Rocket Society, part 2," 70. For the most detailed account of these events, see Neufeld, *The Rocket and the Reich* and *Von Braun*.

21. See Neufeld, *The Rocket and the Reich*, 5–6, 14. See also Neufeld, "The Reichswehr"; Ley's account from *Rockets* (1958), 156; Ley, *Rockets* (1944), 152.

22. "Everything collapsed" from Ley, "What's Wrong," 49; "The beginning" from Ley, Galley Proof for *Inside the Orbit of the Earth*, WLC, 33; "spaceship as the ultimate goal" and "xenophile" from Ley, "End of the Rocket Society, part 2," 70–71.

23. Ley quotations on rockets are from "End of the Rocket Society, part 2," 70–71; Ley, *Grudriß einer Geshichte der Rakete*.

24. See Spohr, "The Final War," *Wonder Stories*, March and April 1932. For Ley's reactions, see Willy Ley, "The Reader Speaks," *Wonder Stories*, August 1932, 282. All Ley quotations in this and the next paragraph are from page 282.

25. On Ley quotations, see previous footnote. For postcard, see Willy Ley to G. Edward Pendray, 30 June 1932, PP, "Ley, Willy 1932."

26. See Walker, *Nazi Science*, 80. On the banning of Lasswitz, see Fischer, *Empire Strikes Out*, introduction.

27. See Winter, *Prelude*, 45. On Valier, see Essers, *Max Valier*. See also Crouch, *Aiming for the Stars*, 50–51.

28. On "swastika armlet," see Ley, *Rockets* (1958), 157. See also Ley, *Rockets* (1944), 153. As stated earlier, Nebel was not a party member.

29. On the Gestapo raid see Neufeld, *The Rocket and the Reich*, 26; Ley, *Rockets* (1958), 160. "Reborn Germany" from Ley, *Rockets* (1944), 153; "No way out" from Ley, *Rockets* (1958), 161. On the decline of the VfR, see also Ley, "What's Wrong," 39–40. On von Braun, see Neufeld, "Wernher von Braun, the SS, and Concentration Camp Labor," 59.

30. The announcement is quoted in Winter, *Prelude*, 48. Ley's "the whole mess," and "the leading spirit" from Willy Ley to G. Edward Pendray, 2 February 1934, PP, box 6, folder 26; Ley to Pendray, 3 April 1934, PP, box 6, folder 26, 1. Ley later claimed that he was only pretending to reorganize the society. See Ley, *Rockets* (1958), 161.

31. See P. E. Cleator to G. Edward Pendray, 5 March 1934, PP, "British Interplanetary Society, P. E. Cleator, correspondence"; Ley to Pendray, 2 February

1934, PP, box 6, folder 26; Ley to Pendray, 23 June 1933, PP, box 5, folder 16, 2. Cleator to Pendray, 5 March 1934, PP, "British Interplanetary Society, P. E. Cleator, correspondence," 1. See Neufeld, *The Rocket and the Reich*, 28.

32. Ley to Pendray, 15 May 1934, PP, box 6, folder 26, 1.

33. See Neufeld, "The Excluded"; Neufeld, *The Rocket and the Reich*, 31. On Brügel's *Manner der Rakete*, see Ley to Pendray, 2 February 1934, PP, box 6, folder 26, 1.

34. See Walker, *Nazi Science*, 7, 12, 71. See also Cornwell, *Hitler's Scientists*, 8.

35. See Cornwell, *Hitler's Scientists*, 191–92.

36. Ley, "Pseudoscience in Naziland," 90; for the quotations that follow, see 90–98. The "Give US Hörbiger" emphasis is in the original article.

37. Ibid., 90–98.

38. On the "mysterious communication," see P. E. Cleator to G. Edward Pendray, 30 October 1934, PP, box 6, folder: "British Interplanetary Society, P. E. Cleator Correspondence, 1934–1935–1936," 1; follow-up response in Cleator to Pendray, 2 November 1934, PP, box 6, "Cleator correspondence," 1.

39. For Pendray's initial invitation, see G. Edward Pendray to Willy Ley, 15 November 1934, PP, box 6; see also Pendray to Raymond T. Geist, 23 November 1934, PP, "Ley, Willy, 1934," 1, box 6, folder 26; Pendray to Ley, 20 November 1934 PP, box 6, folder 27; National Council of Jewish Women to GEP, 21 November 1934. PP, box 6, folder 26.

40. See P. E. Cleator to G. Edward Pendray, 29 November 1934, PP, box 6, folder 27, 1–2; Ley to Pendray, 23 December 1934, PP, box 6, folder 26, 1.

41. See Willy Ley to G. Edward Pendray, 15 December 1934, PP, box 6, folder 26, 1; Ley's plea for funds is in Ley to Pendray, 18 January 1935, PP, box 6, folder 26, 1; on the rejection see Douglas Jenkins to Pendray, 23 January 1935, PP, box 8, folder 1, 1. See also Ley to Pendray, 21 January 1935, PP, box 8, folder 1; Pendray, "Draft," PP, box 8, folder 1; Ley to Pendray, 30 January 1935, PP, "Ley, Willy, 1935," box 8, folder 1, 1.

42. See Willy Ley, "Von Berlin über England nach New York," *Das Neue Fahrzeug*, March 1935, 17, reproduced in the appendix of Ley at al., *Die Starfield Company*; Unknown author, "Rocket Whiz," clipping in WLC; Sharpe and Ordway, *The Rocket Team*, 90.

43. See P. E. Cleator to G. Edward Pendray, 7 February 1935, PP, box 6, "BIS–P. E. Cleator, correspondence" 1. For Ley's correspondence, see Ley to Pendray, 7 February 1935, 1.

44. P. E. Cleator to G. Edward Pendray, 13 February 1935, PP, box 6, 3–9, quotes 5–8.

45. Ley, "Von Berlin über England nach New York."

1. "Beginning" from Ley, "Eight Days," 63–64.

2. On the lack of public interest, see Ley, *Events in Space*, 22. For comment on Goddard see Ley, "Moons and Missiles," 14.

3. Information on Ley's interest in the horseshoe crab is in "Ley, Willy," *Current Biography*, 1941, 513; see also Ley, "The Story of European Rocketry"; Ley and Cleator, "The Rocket Controversy," 19.

4. See Ley, "What's Wrong," 49–51, 145. Quoted in Sharpe and Ordway, *The Rocket Team*, 20.

5. For Ley's letter, see Willy Ley to G. Edward Pendray, 23 June 1935, PP, box 8, folder 1. Information on his stay with the van Dresser family is in a 1951 FBI "Voice of America" background check, obtained through a Freedom of Information Act request. On Pendray's skepticism of space travel, see various public speeches in the Pendray Papers at Princeton.

6. For Pendray's correspondence, see Pendray to Harry Bull, 10 April 1935, PP, box 8, "Pendray—Correspondence Harry Bull—Rockets, 1935." See also Solomon, "Oldies and Oddities," 87.

7. Solomon, "Oldies and Oddities," 86. For the reporter's descriptions, see unknown author, "Rocket Plane Fails to Soar."

8. "Worthwhile" from Ley, "Eight Days," 63.

9. See unknown author, "Rocket Plane Fails to Soar."

10. Quotations from Ley, "Eight Days," 64. See also unknown author, "Two Rocket Plane Flights."

11. See Ley to Pendray, 27 July 1936, PP, "Ley, Willy," 1936, box 8, folder 22. Information from Sandra Ley was obtained through oral history interviews.

12. McNash, "Proposed Altitude Rocket Hops."

13. Ibid.

14. Ley, "The Road to Space Travel, part 1," 73.

15. According to Sam Moskowitz, Ley averaged $12 a week; Moskowitz, "Willy Ley: Forgotten Prophet," 52. Ley's recorded article sales indicate a steady income of at least $27 a week from sold articles. See WLC, box 4, folder 1. See also Cheng, *Astounding Wonder*, 53.

16. Ley (as Robert Willey), "At the Perihelion," 44.

17. Ibid., 47.

18. Ibid., quotations 46–47.

19. Ibid., 49.

20. Ibid., 57.

21. Article sales information from Ley's records of income, WLC, box 4, folder 2.

22. Ley, "Space War," 72. See Ley's "The Dawn of the Conquest of Space"; "Eight Days"; "Visitors from the Void"; "Stations in Space"; "Calling All Martians!"; for other articles, consult the Internet Speculative Fiction Database.

23. See Pendray, "Number One Rocket Man," and "To the Moon via Rocket?"

24. See Ley's articles "Geography for Time Travelers"; "Botanical Invasion"; "Antlantropa"; "The Kitchen of the Future." See also his "The Conquest of the Deep"; "Sea of Mystery"; "Death Under the Sea."

25. Ley, "See Earth First!"

26. All quotations from Ley, "See Earth First!"

27. For a somewhat similar perspective on scientific rebellion, see Pandora, *Rebels within the Ranks*.

28. These public documents were obtained via ancestry.com. Margot listed her husband as W. O. Ley. The address matches Ley's address, according to FBI files. Unverifiable stories are found in Campbell, *John W. Campbell Letters*, vols. 1 and 2; Asimov, *In Memory Yet Green*; Asimov, *In Joy Still Felt*; De Camp, *Time and Chance*.

29. Ley, "The Search for Zero, part 1," 122; other quotations from 122–23.

30. Ibid., 127–30.

31. Ibid., 134–35. Subsequent quotations from Ley, "The Search for Zero, part 2."

32. Ley, "The Search for Zero, part 2," 104.

33. Ley, "The New York World's Fair, 1939"; quotations from 51–52 except time capsule, 56.

34. Mauro, *Twilight at the World of Tomorrow*; Smith, *Making the Modern*; Robert Bennett, "Pop Goes the Future: Cultural Representations of the 1939–1940 New York World's Fair," in Rydell and Schiavo, *Designing Tomorrow*.

35. Nye, *American Technological Sublime*, 223.

36. Unknown author, *Official Guide Book of the New York World's Fair 1939*, 197.

37. Mauro, *Twilight at the World of Tomorrow*, xxiii.

38. For Campbell's comments, see "The Editor," "Fog," *Astounding Science Fiction*, December 1940, 6.

39. Ley (as Robert Willey), "Fog," 81.

40. Ibid., 83–84.

41. Ibid.

42. Ibid., 85–86.

43. Ibid., 88.

44. Ibid., "Everybody" 94, "Jenkins Radio Dome" 88–92.

45. Ibid., 95.

46. Ibid., 96.

47. Ibid., 97.

48. Ibid., 102.

49. Ley, *The Lungfish and the Unicorn*.

50. Ibid., 18.

51. Ibid., 6.

52. Ibid., 5.

53. Ibid., 5–8.

54. Ley, "Science and Truth," 96.

55. Ley, *The Days of Creation,* quotations from page 4 of Ley's introduction.

56. Ibid., 11.

57. Ibid., 42.

58. Ibid., 267.

59. See Hollinger, "The Unity of Knowledge," 212–13. See also Hollinger, *Science, Jews, and Secular Culture.*

60. Hollinger, "The Unity of Knowledge," 214.

61. Benjamin Alpers, *Dictators, Democracy,* roots of Cold War ideology 13.

62. See Hollinger, "Science as a Weapon," 441–42. See also May, "The Morale Code of the Scientist."

CHAPTER 5. THE *PM* YEARS AND THE SCIENCE WRITERS AT WAR

1. Unknown author, "This is PM," "Opinion," inside cover.

2. On "voluntary propagandists," see Baughman, *The Republic of Mass Culture.* For *PM* see unknown author, "War Is at Our Doorstep" 5.

3. Uhl, "Lindbergh Lays Foundation for Fascist Revolt," 8. For full-page pictures of dead children see, for example, "Foreign," *PM,* July 13, 1941, full page.

4. "Presearch" from Willy Ley to Frederik Pohl, 17 June 1940, Frederik Pohl Papers (hereafter cited as FPP), Syracuse University Library, box 7, "Correspondence with Dr. Willy Ley"; "It is, believe me" from Willy Ley to Robert Heinlein, 18 July 1940, "Correspondence, Pre-War, 305J," Robert A. and Virginia Heinlein Archives, University of California, Santa Cruz (hereafter cited as HA).

5. Ley to Pohl, 21 September 1940, FPP, box 7, "Correspondence"; "night force" from Ley to Heinlein, 3 February 1941, HA, "Correspondence, Pre-War, 305J," 1.

6. "Hoax" from Ley to Heinlein, 11 September 1941, HA, box 5, "Correspondence, Pre-War, 305J"; on O'Connor, see Ley to Heinlein, 9 November 1945, HA, box 220-1, "Personal Correspondence, 1943–1971."

7. Ley, "New Weapons Department: The Rocket-Powered Multiple Cannon"; Ley, "The Shooting Range." On *Coast Artillery Journal,* see Ley, "War Rockets of the Past."

8. See Ley, "New Weapons Department: Bat Men."

9. Ley, *Bombs and Bombing,* 51–53, H. G. Wells 55–56.

10. "Morbidly interesting" from Thompson, "Books of the Times"; "vague terror" from K. W., "Fire and Poison From the Air"; "valuable little work" from Goddard, "Review."

11. Sunshine, Ley to Heinlein, 3 February 1941, HA, corr. 305J, 2; Lang, "Sci-Fi Filmmaker's Debt"; see also Ley to Heinlein, 20 July 1941, HA, "Corr Pre-war," 1.

12. For an account of these events see Ley, *Dragons in Amber*, 185–86.

13. Ley to Heinlein, 23 December 1941, HA, box 305, "Correspondence, Pre-War, 305J."

14. See G. Edward Pendray, "To the Members of the American Rocket Society," 15 January 1942, PP, box 11, "American Rocket Society, 1941 General," 1.

15. "The answer is NO!" from Ley, "War Weapons," 9; "present war" and other quotations from Ley, "War Weapons," 10. For a representative sampling of Ley's articles, see *PM*, January 1942.

16. Seversky, *Victory through Air Power*; Ley, *Shells and Shooting*, 11–12.

17. Ley, "Debunking"; for Ley's reply and Quincy Howe, "Publishers of 'Victory through Air Power' Answer Willy Ley," see *PM*, December 11, 1942, 21.

18. For "PS" see Ley to Pohl, 18 January 1943, FPP, box 7, "Correspondence with Dr. Willy Ley"; on poor eyesight see Pohl to Ley, 19 January 1943, FPP, box 7, "Correspondence with Dr. Willy Ley." See also Ley to Heinlein, 10 March 1943, HA, box 220-1, "Personal Correspondence, 1943–1971."

19. "War rockets" from Ley, "War Weapons IX: Rockets," 20. See also Ley, "Rocket Artillery"; Ley, "Nazi's Super Submarine," 12–13. "Many observers" from Ley, "War Weapons: Nazi May Use Gas," 9; and see Ley, "War Weapons: How to Fight Poison Gas," 5.

20. "Table of Contents," *Mechanix Illustrated*, June 1942, 4; for military-related *Astounding Science Fiction* articles, see the Internet Speculative Fiction Database.

21. Ley, *Rockets: The Future of Travel*. See also Ley, "End of the Rocket Society," parts 1 and 2; date of naturalization taken from FBI file, "Request for Investigation," January 4, 1951.

22. Ley, *Rockets* (1944), 3.

23. Ibid., "And as for war rockets" and "I'm going to speak" 4, other quotations 1–3.

24. Ibid., 10.

25. Ibid., judgments on non-Western astronomy 6–7, "astronomical revolution" 4–15, Kepler's *Somnium* 16–19, "things had suddenly" 26.

26. Ibid., moon hoax 28–31, decades of great dreams 48.

27. Ibid., 52–53.

28. Ibid., 122.

29. Ibid., 248–49.

30. See Kaempffert, "Ranging Beyond the Stratosphere." See also "Display Ad 116," *NYT*, May 21, 1944, BR9; "Glimpses of the Moon," *Time*, July 31, 1944, 70–72; Richardson, "Review."

31. On Ley's other articles see, for example, "Bombs vs. Shells," *Mechanix Illustrated*, August 1944, 44–46, 149–50; "Here He Is—the German Who Perfected Hitler's Flying Bomb," *London Daily Mail*, July 5, 1944, 4. For newsprint coverage see "Robots Kill 2,752," *Pittsburgh Sun-Telegraph*, July 6, 1944, front page; "Germans Unleash Giant New Robots," *Washington Post*, July 27, 1944, front page.

32. Ley, "The Future of the Robot Bomb."

33. Sarton, "Preface to Volume XXXIII," 3.

34. "Everything is political" from Montagu, "Review," 298; Montagu, *Man's Most Dangerous Myth*, early quotations from 14–15, "tyranny" 353; last reviewer quotation from Iwanska, "Popular Science and Race."

35. For obituary, see "Waldemar B. Kaempffert Dies; Science Editor of the Times, 79," *New York Times*, November 28, 1956, 35. For other quotations, see Kaempffert, "Science in the Totalitarian State," 435–36. For general works, see Waldemar Kaempffert, *A Popular History of American Invention* (New York: C. Scribner's Sons, 1924); *Science Today and Tomorrow* (New York: Viking, 1945); *Explorations in Science* (New York: Viking, 1953).

36. Kaempffert, "Science in the Totalitarian State," 434–36.

37. Ibid., 441.

38. See, for example, Waldemar Kaempffert, "The Soviet Way," *New York Times*, November 7, 1943, E11. See also "What Are Scientists Doing?" in *America Organizes to Win the War: A Handbook on the American War Effort*, ed. E. M. Hunt (New York: Harcourt, Brace, and Company, 1942); Waldemar Kaempffert, "American Science Enlists," *NYT*, November 2, 1941, SM3; *Should the Government Support Science?* (New York: Public Affairs Committee, 1946).

39. O'Neill, "Writer Charges U.S. with Curb on Science."

40. Lindeman, "Introduction," ix–x.

41. Ibid.

42. Jewett, *Science, Democracy*.

CHAPTER 6. AN ENGINEER'S POSTWAR DREAMS

1. Public charter records indicate that Burke Aircraft Corporation was incorporated on June 13, 1944.

2. Cleator, "Tribute to Willy Ley."

3. Ley, "Notes on Weapons: V-2, V-1, Me-163," *Technology Review*, January 1945, 169. See also *Atlanta Daily World*, November 14, 1944, 4; November 28, 1944, 4; December 5, 12, 19, and 27, 1944.

4. For related coverage, see Botsford, "The Rockets Rain Down"; Ley, "V-2 Rocket Cargo Ship," 100. See also Ward, *Dr. Space*, 43. Source is listed as "A. V. 'Val' Cleaver, letter, February 16, 1972, 'X = 60 and Counting.'"

5. For "full story," see Ley, "V-2 Rocket Cargo Ship," 104; Oberth "department head" comment 110, other quotations 121–22.

6. See *The Breeze* 2, no. 4, May 10, 1945, posted online by NOAA. See also Willy Ley to Robert Heinlein, 21 June 1945, HA, box 220-1, "Personal Correspondence, 1943–1971," 1.

7. Ley to Heinlein, 21 June 1945, 1.

8. On earnings and meteorological rockets, see Ley to Heinlein, 3 August 1945, HA, box 220-1, "Personal Correspondence, 1943–1971," 1.

9. Ley to Heinlein, 4 July 1945, HA, box 220-1, "Personal Correspondence, 1943–1971," 1.

10. Ley to Heinlein, 3 August 1945, HA, box 220-1, "Personal Correspondence, 1943–1971," 2.

11. Conversation with Laning and "Moon Messenger" are from Ley to Heinlein, 3 August 1945, 1; response from Heinlein to Ley, 15 October 1945, 1; improved relations with Laning and welding seams from Ley to Heinlein, 9 November 1945, 2, all in HA, box 220-1, "Personal Correspondence, 1943–1971."

12. See Ley, "Inside the Atom"; Shapiro, "Man and the Atom."

13. Ley to Heinlein, 9 November 1945, HA, box 220-1, "Personal Correspondence, 1943–1971," 2–4.

14. Ley, "Peace or Else!" 45–46, "so incredible" 47. On the "gentlest possible application" and subsequent quotations, see page 72. See also Ley, "Atomic Engines for Peace" and "Atomic Medicine."

15. This account is presented in Ley, "Improving upon the V-2," Ley's reaction to lecture 100–1, V-2 exhibit 102–3.

16. See Ley to Heinlein, 7 May 1946, HA, box 220-3, "Personal Correspondence, 1943–1971," 1–2.

17. Ley to Heinlein, 6 June 1946, HA, box 220-3, "Personal Correspondence, 1943–1971," 1. Earlier quotations from Ley to Heinlein, 20 May 1946, HA, box 220-3, "Personal Correspondence, 1943–1971," 1.

18. For "our friend the Navy," "The official request," and von Braun, see Ley to Heinlein, 29 August 1945, HA, box 220-3, "Personal Correspondence, 1943–1971," 1.

19. Ley to Heinlein, 16 September 1946, HA, box 220-3, "Personal Correspondence, 1943–1971," 2.

20. Pendray, "Next Stop the Moon."

21. Pendray, "The Reaction Engine," 70. For a richer discussion of Pendray's early activities, see John Cheng's *Astounding Wonder*, 251–300. On Pendray's earlier articles, see the Pendray Papers at Princeton. For quotation on "one of the most successful weapons of modern warfare," see Pendray, "Skyrockets Grow Up," 4–5.

22. Pendray, *The Coming Age of Rocket Power*; Ley, "Review." For a critical review that mentions "numerous other persons," see unknown author, "Turns with a Bookworm," WLC. See also Marks, "Jet Propulsion Theories," 240–41.

23. Clarke, *The Exploration of Space*; "For myself" from Pendray, "Age of Rocket Power," 159. For reviewer's comment on Opel's publicity, see Rufus Oldenburger, "All about the Rockets: Layman's Book Tells Potentialities," *Chicago Sun Book Week*, June 3, 1945, clipping in Pendray Papers.

24. On Goddard's influence on Oberth, see Pendray, "The Persistent Man," 54. This article is extracted from Pendray's *The Coming Age of Rocket Power*. For obituary, see Pendray, "Robert H. Goddard."

25. See Willy Ley to Herbert Schaefer, Willy Ley Collection, box 1, folder 1: "Correspondence, 1945–1949," 1. Translation taken from Neufeld, *Von Braun*, 232.

26. For "He [Ley]" quotation, see Laning to Heinlein, 6 December 1946, HA, box 017, Opus 052, "A Spaceship Navy aka Flight to the Future," 1; for reactions to the von Braun reunion, see Heinlein to Laning, 12 January 1947, HA, box 017, Opus 052, 1; Laning to Heinlein, 24 January 1947, HA, box 017, Opus 052, 1; on the discontinuation of sponsorship, see Heinlein to Laning, 28 January 1947, HA, box 017, Opus 052, 1.

27. Ley, *Rockets and Space Travel* (1947), foreword 1, other quotations 2–4. For earlier passage on "modern war rockets," see Ley, *Rockets* (1944), 78.

28. Ley, *Rockets* (1947), 207.

29. Ibid., 221. On selling point, see "Rockets to the Moon," *Newsweek*, March 24, 1947, 64.

30. Neufeld, "Creating a Memory," 72–73.

31. Ley, *Rockets* (1947), 229.

32. Gardner, "Through Pathless Realms." See also Ley, *Rockets* (1947), 314.

33. See *Coast Artillery Journal*, March–April 1948. For review, see "Display Ad 220," *New York Times*, May 25, 1947, BR31. See also Gardner, "Through Pathless Realms."

34. On moral outrage, see Neufeld, *Von Braun*, 235.

35. Ley continued to contribute to science fiction pulps, albeit less frequently in the late 1940s. For *Astounding Science Fiction*, see Ley's "Improving upon the V-2"; "Pseudoscience in Naziland"; "Push for Pushbutton Warfare"; "Science and 'Truth.'" For *Startling Stories* and other pulps, consult the Internet Speculative Fiction Database. For representative issue with I. Bernard Cohen, see *Technology Review*, January 1946.

36. Bush, "Science, Strength, and Stability," 553. On the scientific way see Bush, "The Scientific Way," 464.

37. Oppenheimer, "Physics in the Contemporary World," 231; Haslam, "World Energy and World Peace," 493.

38. Killian, "Our Shared Convictions," 503–4, 550. Compton, "Education for Peace," 501–3; Keenan, "Education for Freedom." See also Phillips, "What Is Democracy?"

39. Killian, "Academic Freedom and Communism."

40. Bush, *Modern Arms and Free Men*, 200–1. See also Zirkle, *The Death of a Science in Russia*; Christman, *Soviet Science*; Joravsky, "Soviet Views on the History of Science"; Sax, "Review," 238.

41. Lundberg, *Can Science Save Us?*, 2. Lundberg disagreed with some of his colleagues' Cold War rhetoric; see 45–46. "Unified method of attack" 13; "magical thinking" and "prescientific modes" 4–5; laymen 11; "faith" 115.

42. Conant, *On Understanding Science*, "unified, coherent culture" 19. See also Sherwood, "Science in Education," 98.

43. Ley, "Rockets in Battle," 95. In *Technology Review*, see also Willy Ley, "The Delayed Invention," February 1950, 207–10, 230; "The Unchangeable Ship," January 1951, 147–49, 174.

44. "Alumni Day—1948," *Technology Review*, July 1948, 507. For representative articles see Willy Ley's "The Two-Thumbed 'Teddy Bear,'" *Natural History*, September 1948, 328–32; "The Story of the Fish Anguilla," *Natural History*, February 1949, 82–85, 93; "The Story of the Milu," *Natural History*, October 1949, 373–77; "Do Prehistoric Monsters Still Exist?" *Mechanix Illustrated*, February 1949, 79–83, 140–44.

45. Ley, *The Lungfish, the Dodo, and the Unicorn*. For reviews, see Starrett, "Books Alive"; Leiber, "Some Strange Animals"; Unknown author, "A Line O' Type or Two"; Prescott, "Books of the Times."

46. Mann, "Fancy and Fact," 358; Zirkle, "Review" (1949); George Sarton to Willy Ley, 3 March 1951, WLC, box 1, folder 5.

47. See biographical draft in Pendray Papers, box 12, folder 18. See also Neufeld, *Von Braun*, 238–39; Ley, "Want a Trip to the Moon?"; Ley to Heinlein, 26 July 1948, HA, box 220-4, "Personal Correspondence, 1943–1971."

48. See Miller and Durant, *The Art of Chesley Bonestell*, chapter 3.

49. Ley, *The Conquest of Space*, quotations from "Introduction: Mostly About Chesley Bonestell," 10–11. See also McCurdy, *Space and the American Imagination*, 33.

50. Ley, *The Conquest of Space*, 17–18, countdown 20, classroom and professor 23–24, 31–32.

51. Ibid., 48, "small wonder" 43, "nearly within reach" 46.

52. Ibid., equatorial launch 49–51, quotation on earth 52, "We'll never know" 71.

53. On copies sold, see Ley to von Braun, 8 February 1950, Wernher von Braun Papers, U.S. Space and Rocket Center, Huntsville, Alabama, 406–8. "Hold your hats, everyone!" from Blakesley, "We're Off to the Moon!," H4; see also Pfeiffer, "Round-Trip Ticket to the Moon," BR 29; Unknown author, "It Is X-Hour Minus 5."

54. Krutch, "Responsible Fantasy"; Prescott, "Books of the Times" (1949), 27.

See also Pettitt, "Review"; Gardner, "Bridge to the Moon," 71; Heinlein, "Baedeker of the Solar System," 9.

55. Ley to Heinlein, 14 March 1950, and Heinlein to Ley, 17 April 1950, both in HA, box 306-09, "General Correspondence, 1948–1951."

56. Ley to von Braun, 8 February 1950, Wernher von Braun Papers, Huntsville, 406–8.

Chapter 7. The Tom Corbett Years

1. Lieber, "Whispering Willy."

2. For *Los Angeles Times*, see Scheuer, "Hollywood Will Reach Moon First."

3. Willy Ley to Robert Heinlein, 28 June 1949, 1; Heinlein to Ley, 30 June 1949, 1; Ley to Heinlein, 1 October 1949, 2, all in HA, box 306-09, "General Correspondence, 1948–1951, section 2."

4. *Tom Corbett, Space Cadet*, Episode 1: "The Solar Guard Academy," parts 1, 2, and 3, CBS, 1950, quotation at 4:30–6:15.

5. Ibid., 12:11–12:42.

6. Unknown author, "Hi-yo, Tom Corbett!"

7. Ley to Heinlein, 4 May 1951, HA, box 306-09, "General Correspondence, 1948–1951, section 2," 2. Interview with Willy Ley, *Night-Beat*, with Mike Wallace, July 1957, 18:32, available at the Willy Ley Collection, University of Alabama–Huntsville; special thanks are due to archivist Anne Coleman, who provided an audio recording of this interview.

8. Thomas, "Frankie Thomas on Tom Corbett, Space Cadet." See also Ley to Heinlein, 4 May 1951; Unknown author, "Hi-yo, Tom Corbett!"

9. Ley to Heinlein, 4 May 1951.

10. Ley, *Rockets, Missiles and Space Travel* (1951), Peenemünde 197. For Ley's comments to a reporter, see Breit, "Talk with Willy Ley." For Neufeld's description, see "Creating a Memory," 73.

11. "Serious" from Gannet, "Books and Things," clipping in WLC. See also Krutch, "Review"; Gibbons, "All Aboard for Trip to Moon or Some Such"; Unknown author, "Science Fiction Bookshelf." On the book club, see "Display Ad 971," *New York Times*, September 30, 1951, 178.

12. On Langley Field, see Ley to Heinlein, 25 November 1951, HA, box 308a-3, "General Correspondence, 1952," 1. See also "Authority on Space Ships Will Speak Here Tuesday," *Garden City News*, April 19, 1952; *Mike and Camera*, March 1951, 9, clipping in WLC; "TV Blamed Anew as Rival of Books," *New York Times*, November 18, 1950, 18; "World View Taken by Young Readers," *NYT*, November 19, 1950, 79.

13. Ley, "You'll Live to See a Spaceship," 9; see also Ley, "Out of this World by Spaceship"; Breit, "Talk with Willy Ley."

14. Ley, *Dragons in Amber*.

15. Ibid., foreword, vii.

16. Ibid., amber 20–22, "sinners" 75, "bones" 98, "restrained" 29; "nonsense" 41.

17. Prescott, "Books of the Times," 1951; see also Burns, *Invasion of the Mind Snatchers*, 21; Derleth, "A Scientific Adventure in Romance"; Huxley, "The Romance of Nature," *Observer*, October 21, 1951, clipping in WLC.

18. Jackson, "Bookman's Notebook"; Swanson, "Review," 77.

19. See Ley, *Rockets* (1957), 330.

20. See McCurdy, *Space and the American Imagination*, 35: original source listed as Willy Ley, letter of invitation, First Annual Symposium on Space Travel, 1951, American Museum of Natural History, Hayden Planetarium Library, New York, quoted in McCurdy, 35–36. See also Marché, *Theaters of Time and Space*, 99–100.

21. See Fred L. Whipple, "Recollections of Pre-Sputnik Days," in Ordway and Liebermann, *Blueprint for Space*, 129, quoted in McCurdy, *Space and the American Imagination*, 38. See also Ward, *Dr. Space*, 87. Source quoted as Fred L. Whipple, letter, March 2, 1972, "X + 60 and Counting." See also Ryan, *Across the Space Frontier* and *Conquest of the Moon*.

22. Quoted in McCurdy, *Space and the American Imagination*, 40. See *Collier's*, March 22, 1952.

23. Ward, *Dr. Space*, "space gospel" 87–88, Ryan 88. Source listed as Milton W. Rosen, letter, February 16, 1972, "X + 60 and Counting." See also Neufeld, *Von Braun*, 252, 259.

24. Quoted in Ward, *Dr. Space*, 89. Source listed as Milton W. Rosen, letter, February 16, 1972.

25. Sandra Ley's comments are taken from a 2014 oral history interview with the author.

26. De Camp and Ley, *Lands Beyond*.

27. Ibid., 3–4.

28. Ibid., 5–7.

29. Ibid., 7–8.

30. Ibid., debunking 17–18, "occultist" 19, "commentary and cross-commentary 22.

31. Ibid., Ley on Plato 22–26, Ley's conclusion 43.

32. Ibid., "masterpieces" 113, "cult mind" 171, last quotation 320–21.

33. Poore, "Books of the Times." For other reviews see Amrine, "Those Cities of Gold"; for "readers ride" see unknown author, "Readers Ride to Mythical Destinations," *Ft. Wayne, Ind. News-Sentinel*, June 21, 1952, clipping in WLC; on theatricality see Ruth D. Ray, "Far Away Lands," *Hartford* (Conn.) *Courant*, July 6, 1952, clipping in WLC. Other reviews include Joan Stanford Bishop, "Man's Cupidities and Dreams in Mythology of Geography," *Bridgeport* (Conn.) *Post*,

July 13, 1952; Glenn Negley, "Wily Wilds," *Saturday Review of Literature*, July 26, 1952, clipping in WLC.

34. *Galaxy* Publishing Corp. contract, 1 July 1952, WLC, box 4, folder 3.

35. See "For Your Information," *Galaxy*, September 1952, 49–59. Cheng, *Astounding Wonder*, 77.

36. Ley, "When Will Worlds Collide?"; Ley, "Space Travel by 1960," 90–91.

37. Ley, "The World of October 2052."

38. For representative space-related *Galaxy* articles during these years, see Ley, "Mars," November 1952; "The Birth of the Space Station," parts 1 and 2, April and May 1953; "Mail by Rocket," August 1954; "The Orbital (Unmanned) Satellite Vehicle," July 1955; "Unveiling the Mystery Planet," September 1955; "The How of Space Travel," October 1955. On editor's introduction, see "Any Questions?" *Galaxy*, April 1953, 63.

39. Quoted in McCurdy, *Space and the American Imagination*, 41. Source Smith, "They're Following Our Script," 54.

40. "Write your own ticket" from Smith, "They're Following Our Script," 56; Ley's personal memories can be found in Ley, "The How of Space Travel," 62–63.

41. Smith, "They're Following Our Script," 56–57. See also Ley, *Rockets* (1957), 331.

42. McCurdy, *Space and the American Imagination*, 47; Neufeld, *Von Braun*, 286.

43. Ley, "The How of Space Travel," 64.

44. Asimov quoted in Henry Petroski, "Engineers' Dreams," *American Scientist* 85 (July–August 1997): 310, original source listed as Asimov, *Asimov's Biographical Encyclopedia*; Ley, *Engineers' Dreams* (1964 edition), 271.

45. Ley, *Engineers' Dreams*, 11.

46. Ibid., "these factors" and "what will happen" 11–12, "the word fantastic" 16; "tapping the Earth's interior" 90–92, "comforting fact" 93, "inexhaustible source" 162.

47. Ibid., 135, "In the end" 124.

48. Ibid., "result" 148, "political situation" and "united Europe" 153.

49. For commentary on the insensitivities, see Henry Petroski, "Engineers' Dreams," 312.

50. Ley, *Salamanders and Other Wonders*, x.

51. Ibid., 93, yeti 107.

52. Ibid., new discoveries of strange creatures 147–48, "through centuries" 219.

53. Zirkle, "Review" (1956).

54. On classification of the book, see White, "Review."

55. Cagle, "Review"; Monroe, "Pleasant Dip into Scholarship," clipping in WLC; Unknown author, "Books in Brief"; Gardner, "Review"; Krutch, "Naturally

It's Strange." For "prehistoric animals or spaceships" quotation see Collins, "What Happened to the Dodo Bird."

56. Derleth, "The Romantic Naturalist"; Walton, "Some Old Secrets of Biology," *Virginian Pilot* (Norfolk), August 28, 1955, clipping in WLC; Ball, "Romance and Common Sense," *Times Dispatch* (Richmond, Va.), September 3, 1955, clipping in WLC.

57. Unknown author, "TWA Offering 'Flight to the Moon' Feature."

58. See Strodder, *Disney Encyclopedia*, 37, 390.

59. "Satellite Called Stride in Space; Experts Aware of Its Possibility," *New York Times*, July 30, 1955, 9.

60. *Face the Nation, Volume 1, 1954–1955*, 324.

61. Ibid.

62. Ibid., 325–26, scientific value of a trip to the moon 331.

63. "Events Today," *NYT*, January 4, 1956, 20; Ley to Heinlein, 19 March 1956, HA, corr-309-3, 1.

Chapter 8. The *Sputnik* Challenge

1. Interview with Willy Ley, *Night-Beat*, with Mike Wallace, July, 1957, available at the Willy Ley Collection, University of Alabama–Huntsville; recording provided by archivist Anne Coleman.

2. For a debunking of the extent of the "panic," see McQuaid, "Sputnik Reconsidered."

3. Ley and von Braun, *The Exploration of Mars*, 134.

4. Ibid., "genus homo" 85, "step-by-step" 97.

5. Pfeiffer, "Traveling In Space." See also unknown author, "Reading for Pleasure."

6. Ley, "Rockets into Space," 65, 69.

7. Ibid., 69.

8. For other articles, consult the Internet Speculative Fiction Database. For quotations, see Ley, "Between Us and Space Travel," re-entry problem 102–3.

9. See "On Radio," *New York Times*, June 28, 1956, 59, and July 20, 1956, 29; see also "Events Today," *NYT*, January 4, 1956, 20. On lectures see *P.S.*, vol. 2, no. 4 (May 1956): 1; *Scanner*, vol. 6, no. 4 (February 1956), cover. For navy lecture see "Ley Contends New Vanguard Satellite Could Have Been Built 50 Years Ago," *NYT*, April 22, 1956, 73.

10. Ley, *Rockets, Missiles, and Space Travel*, rev. ed. (1957). See also Launius, *Frontiers of Space Exploration*, 190.

11. Ley, *Rockets* (1957), 323, "What holds it up?" 327–28. For *Technology Review* article, see next note.

12. Ley, "The Satellite Rocket," 93.

13. Ley, *Rockets* (1957), 330.

14. See Walters, "TV Ticker." See also Fink, "Space Expert Helps Kids to Learn About It." For quotations, see "Display Ad 191"; Ley, *Man-Made Satellites.*

15. Ley, *Man-Made Satellites*, 7–9.

16. Ibid., 44.

17. On popularity, see unknown author, "Scientist Catapulted to Fame," *Athena Press* (Oregon), December 11, 1958, clipping in WLC; Ley, *Space Pilots*, "the people" and "Some of you" 8–9, training 17.

18. Ley, *Space Pilots*, 41, 44. For other books see Ley, *Space Stations* (1958) and *Space Travel* (1958).

19. Interview with Willy Ley, *Night-Beat*, with Mike Wallace, July 1957.

20. See Ley, *Rockets, Missiles, and Men in Space* (1968), 330–32.

21. For a good introduction to the historiography, see Launius et. al., *Reconsidering Sputnik*; see also McDougall, *The Heavens and the Earth.*

22. Bronowski, *Science and Human Values*, 79. See also the popular books of physicist George Gamow.

23. Bronowski, *Science and Human Values*, 80; Bush, *Modern Arms and Free Men*, 200.

24. Rudolph, *Scientists in the Classroom*, 29, 52. See also Jewett, *Science, Democracy, and the American University.*

25. Unknown author, "Scientist Catapulted to Fame," clipping in WLC; Ley to Heinlein, 26 October 1957, HA, corr. 310-3.

26. Del Rey, "Credo"; on headlines see "German Gives Russia Full Credit for Satellite," *Los Angeles Times*, October 7, 1957, 10, 18; on debunking the myth of Russian "Germans" see, for example, Willy Ley, "Myth of German Experts Debunked," *LAT*, May 18, 1958, 27; "What German Rocket Experts Did in Russia," *Philadelphia Bulletin*, May 21, 1958, clipping in WLC. Lastly, see "A Recovery of Dog Called Possible," *New York Times*, November 4, 1957, 8; Wayne Phillips, "New Launching Seen Near," *NYT*, October 7, 1957, 1.

27. For the announcement, see *Chicago Sun-Times*, November 17, 1957, 14. Many of the following quotations come from the *Los Angeles Times* (hereafter cited as *LAT*), solely due to their accessibility through Proquest. See also "Special Willy Ley Issue," *Monogram*, February 1959, 1; "Rocket Expert's Space Series Starts Today," *LAT*, November 10, 1957, 1; Ley, *Satellites, Rockets, and Outer Space.*

28. See Willy Ley, "What Will Invaders from Space Look Like?" *LAT*, November 10, 1957, K17; "Moons and Missiles: Von Braun Was Rocket Pioneer," *LAT*, November 14, 1957, 2; "Moons and Missiles: Space Science Pleas Ignored," *LAT*, November 18, 1957, 14; on boldest prediction see Ley, "Commerce of the Skies," in *Satellites, Rockets, and Outer Space*, 107; on "Operation Outer Space" see Ley, "Moons and Missiles: Space Venture Steps Outlined," *LAT*, November 19, 1957, 23; "space travel will follow" from Ley, "A Two-Year Look into Space,"

Rotarian, October 1958, 56; Mary Middleton, "Women Plan May Benefit for Hospital," *Chicago Daily Tribune*, March 25, 1958, A4.

29. "Television Programs," *New York Times*, November 24, 1957, 148; Unknown author, "5 Youths Define Race for Space"; Willy Ley, "For Your Information: IGY Roundup," *Galaxy*, July 1958, 56.

30. Ley, "Some Implications of the Sputniki," 515–16; Ley, "Soviets Seen Far Ahead in IRBMs."

31. See Neufeld, *Von Braun*, 312.

32. See "Scholastic Press Meets This Week," *New York Times*, March 9, 1958, 49; "TV Satellite for Weathermen Seen," *LAT*, February 23, 1958, A1; on "new communications era" see "How Satellites Will Bring New Communication Era," *Chicago Sun-Times*, December 21, 1958, clipping in WLC; "A Shot to Moon This Year? Here Are 2 Methods," *CS-T*, March 2, 1958, page unknown. See also Ley, "1958 May Be Year of Cluttered Sky," *LAT*, March 2, 1958, 30.

33. Willy Ley, "First Moon Trip by 1965 Possible," *LAT*, March 16, 1958, 22. See also "Moon Trip Forecast Within Seven Years," *LAT*, March 21, 1958, 20; "Moon Should Be First Target," *CS-T*, March 8, 1959; Ley, "Moons and Missiles: Space Venture Steps Outlined," *LAT*, November 19, 1957, 23. On red, white, and blue space station, see "Expert Expects Manned Space Station by '65," *LAT*, February 24, 1959, D19.

34. On ridicule, see Ley, "Idea of Rockets Mocked, Hailed," *LAT*, November 13, 1957, 2; On Ley and Pendray feud, see correspondence in "GEP Correspondence with United Feature Syndicate," PP, box 64.

35. Ley, Willy, "America Snubbed Her Scientists," *Birmingham Post and Gazette*, November 23, 1957, page unknown; Ley, "Moons and Missiles: Space Science Pleas Ignored," *LAT*, November 18, 1957, 14.

36. See Ley, "Moons and Missiles: Space Science Pleas Ignored," *LAT*, November 18, 1957, 14; see also Willy Ley, "Willy Ley Analyzes Aspects of Space Jaunt," *LAT*, January 7, 1958, 12.

37. Willy Ley, "Russia Provides Rocketry Books," *LAT*, April 20, 1958, 26.

38. See, for example, Willy Ley, "Solid Rocket Fuel Seen Taking Lead," *LAT*, February 2, 1958, 22; quotations from Willy Ley, "Conquest of Space Vital for Nation," *LAT*, January 5, 1958, 24.

39. "We are lagging" from Ley, "Expert Expects . . . ," *LAT*, February 24, 1958, D19.

40. See Parry, *Russia's Rockets and Missiles*, 10–11; Sherman, "Willy Ley Appalled by U.S. Rocket Lag."

41. Ley, "Who Owns Space?" See also Ley, "Sputnik III Is Dual Type of Satellite," *LAT*, May 25, 1958, A22. For comments on Project Vanguard, see Willy Ley, "Vanguard Career Is Ending; New Missile May Use Name," *Chicago Sun-Times*, October 12, 1958, clipping in WLC. On Russian risks to cosmonauts see "Expect Reds to Put Man in Orbit First," *Daily Defender*, February 25, 1959, 2.

42. "Space Exposition Is Staged in Store," *New York Times*, February 5, 1958, 11.

43. "Special Willy Ley Issue," *Monogram*, February 1959, 1–3.

44. Ibid., 3. On "space conscious America" see "Willy Ley Space Models," *Monogram*, May 1959, 2.

45. "Special Willy Ley Issue," *Monogram*, February 1959, 4.

46. See *Space Age News Letter*, ed. Willy Ley, October 1959, 1. See also "Willy Ley Describes Your Flight to the Moon," *Sunday Midwest*, July 12, 1959, 9; "best way to learn" from Sherman, "Willy Ley Appalled by U.S. Rocket Lag."

47. "Monogram Space Models Lead in Hobby Kit Popularity!" in *Space Age News Letter*, 3.

48. Walt Disney Productions, *Man in Space*, 35. See also Walt Disney Productions, *Tomorrow the Moon* and *Man and Weather Satellites*.

49. Walt Disney Productions, *Mars and Beyond*, "the first Mars ship" 11, "fleet of ships" 20, colonization 48.

50. Interview with Willy Ley, *Night-Beat*, with Mike Wallace, July 1957.

51. Unknown author, "Scientist Catapulted to Fame," clipping in WLC.

52. Ley and von Braun, *The Conquest of Space* (recording), accessible at archive.org.

Chapter 9. Scholarly Twilight

1. Moskowitz, "Willy Ley in the U.S.A., part III," "dominant position" and "running" 18, other information 19–20.

2. See Ley, *The Borders of Mathematics*; Ley's activities also included editing and translating *Otto Hahn: A Scientific Autobiography* (New York: Charles Scribner's Sons, 1966). Additionally, he contributed to an English translation of Kant's *Cosmogony* (New York: Greenwood, 1968).

3. In 1960 Ley appeared on the television program *Sun and Substance*. Other television appearances include *At Your Beck and Call*, August 11, 1961, and *Camera Three*, January 10, 1965. His radio interviews are too numerous to cite. See Thonssen, *Representative American Speeches*, 115–30.

4. See Ley, "Spaceways: A Fortress in Space."

5. Ley, "Are We Going to Build a Space Station?" See also Ley, "The End of the Jet Age," October 1962, 73–81; "Sounding Rockets and Geoprobes," April 1963, 88–101; "Anyone Else for Space?" June 1964, 110–28; for other articles, consult the Internet Speculative Fiction Database.

6. Ley, "Points of View: It's Not Science versus the Humanities." See also Snow, *The Two Cultures*.

7. Ley, "Points of View"; Willy Ley, Manuscript, *Not Beyond Understanding*, WLC, box 7, folder 1.

8. Ley, "The Space Age and Education."

9. Ley, *The Poles*.

10. Ibid., 9, 31.

11. Ibid., 129–30.

12. Ibid., 136.

13. Ibid., 151–52.

14. Ley, *Watchers of the Skies*, xi.

15. Ibid., xii.

16. Ibid., three conditions 4–6, Chaldeans 10, Egyptian and Jewish astronomy 19–21.

17. Ibid., 30–31.

18. Ibid., 42–43.

19. Ibid., Ptolemy 48, subsequent quotations 31.

20. Ibid., Martin Luther and Copernicus 61, Scientific Revolution 70.

21. Ibid., "inevitable collapse" 82, Brahe 90, Kepler 93.

22. Ibid., 93.

23. Ibid., Galileo 105–6 and 109, correspondence 113.

24. Ibid., 119, 121–23.

25. Ibid., official silencing 125, "churchmen and public" 126, "immediate success" 128; subsequent quotations 129.

26. Ibid., 130–31.

27. Ibid., 504.

28. Moskowitz, "Willy Ley in the U.S.A., part III," 19–20.

29. Davenport, "Watchers of the Skies"; Gingerich, "Watchers of the Skies."

30. Cohen, "Guide for Space Travel"; Schwartz, "Books of the Times."

31. Hellman, "Review," 377.

32. Kuhn, *The Copernican Revolution*, viii.

33. Kuhn, *The Structure of Scientific Revolutions*, 1–3, ingrained ideology 138.

34. Carlson, "Review." James Kip Finch made similar points in a review in the winter 1962 issue of *Technology and Culture*.

35. Asimov, *A Short History of Biology*; Mayerhöfer, "Review," 369; Asimov, *Asimov's Biographical Encyclopedia of Science and Technology*; Holmes, "Review."

36. Hahn, *Otto Hahn: A Scientific Autobiography*, trans. by Willy Ley (New York: Charles Scribner's Sons, 1966). For sample review, see Maddock, "Review," 583. Many reviews of James Watson's *The Double Helix* were critical.

37. Topsell, *History of Four-Footed Beasts and Serpents*; Stannard, "Review," 335; Koestler, *The Sleepwalkers*. For a scathing review see Santillana and Drake, "Arthur Koestler and His Sleepwalkers."

38. Reingold, "Review," 275. See Gilman, *Science: U.S.A.*

39. Chambers, "History of Science on the Silver Screen," 496–97.

40. Joravsky, "Review," 348; Joravsky, *Soviet Marxism and Natural Science*. See also the Autumn and Winter 1961 issues of *Technology and Culture*. In particular see Joravsky, "The History of Technology in Soviet Russia and Marxist Doctrine."

41. Agassi, *Towards a Historiography of Science*. On "non-scholarly" comments, see Mills, "Review." On the dichotomy, see Cannon, "Review," 391. See also J. Brookes Spencer's review of D.K.C. MacDonald's *Faraday, Maxwell, and Kelvin* (Garden City, New York: Doubleday, 1964) in *Isis* 56, no. 3 (Autumn 1965): 392–93; Garland F. Allen's review of James Watson's *The Double Helix* (1968) in *Isis* 59, no. 4 (Winter 1968): 464–66.

42. Mosse, "Review"; Sherwood, *The Alchemists*; on initial reception see Miles, "Review."

43. Hall and Hall, *A Brief History of Science*. See also A. Rupert Hall, *The Scientific Revolution*, xii; Marie Boas Hall, *The Scientific Renaissance*, 168; Hall, *From Galileo to Newton*; Cohen, "Review," 240.

44. Hoyle, *Of Men and Galaxies*, quoted in Cannon, "Review," 249.

45. See Ley, ed., *Harnessing Space* (1963), 16. See also Ley, *Inside the Orbit of the Earth* (1968). For juvenile and general scientific works see Ley, *The Meteorite Craters* (1968); *The Discovery of the Elements* (1968); *Visitors from Afar: The Comets* (1969); *Gas Planets* (1969).

46. Ley, *Beyond the Solar System*, 23.

47. Ley, *Our Work in Space*; *Events in Space*. See Ley, *Missiles, Moonprobes, and Megaparsecs*; *Ranger to the Moon*; *Mariner IV to Mars*; "Cold War in Space." See also Ley, "The Next Five Years in Space."

48. Ley, *Rockets, Missiles, and Men in Space*, vii.

49. Ibid., astronauts 374, 383–86.

50. Ibid., deaths of astronauts 400, exploration 418.

51. On encyclopedia, see unknown author, "Books. Space Flights"; Earley, "Beyond the Earth"; Emme, "Space, Past, Future," 874. For *Houston Post* quotation see "Display Ad 378," "Display Ad 330."

52. Ley, *Dawn of Zoology*, viii.

53. Ibid., 5. Krutch, "Introduction," in Ley, *Dawn of Zoology*, vii. Quotation on casual observer 6, comments on prehistory 9; Greeks 10.

54. Ibid., 18.

55. Ibid., Aristotle 10, Plato 27, Pliny the Elder 43–47, 54, "plethora" 51.

56. Ibid., *Physiologus* 75.

57. Ibid., 76–79.

58. Ibid., 81–85.

59. Ibid., 86–88.

60. Yarber, "Review," 272; Stannard, "Review"; Brown, "Review," 335.

61. For a critique and historical framing, see Golinski, *Making Natural Knowledge*, 2.

62. Keay Davidson, "Why Science Writers Should Forget Carl Sagan and Read Thomas Kuhn: On the Troubled Conscience of a Journalist," in Doel and Söderqvist, eds., *The Historiography of Contemporary Science, Technology, and Medicine*, 15–30, quotations 21.

63. Golinski, *Making Natural Knowledge*, 9; Davidson, "Why Science Writers," 9; John R. G. Turner, "The History of Science and the Working Scientist," in Olby et al., *Companion to the History of Modern Science*, 30.

64. Reisch, *How the Cold War Transformed the Philosophy of Science*, 3.

65. Ibid., 6.

EPILOGUE: FIRST CITIZEN ON THE MOON

1. See "Personal Materials, Moon Book Project," Willy Ley Collection, box 6, folder 9.

2. Unknown author, "Willy Ley, 62, Space Travel Expert, Dies"; Asimov, *In Joy Still Felt*, 494; Sullivan, "Willy Ley, Prolific Science Writer, Is Dead at 62"; Lang, "Sci-Fi Film-maker's Debt to Rocket Man Willy Ley."

3. Bonestell, "Tribute to Willy Ley," 409. See also Cleator, "Tribute to Willy Ley."

4. Del Rey, "Credo," 151–53.

5. Ibid., 156–57.

6. Ley, "The Orbit of Explorer-I." See also "Ley, Willy," in Anonymous, *World Authors*, 1553.

7. Von Braun, "foreword," in *Beyond the Solar System*, iii. See also Geppert, "Introduction," in *Imagining Outer Space*.

8. For an interesting and retrospective analysis, see Benjamin, *Rocket Dreams*, 2–4.

9. Kilgore, *Astrofuturism*, 74–76.

10. "We Are the Explorers," publicity video, NASA, 2012.

11. Interview with Willy Ley, *Night-Beat*, with Mike Wallace, July 1957, 12:55–13:00.

12. Stephen J. Pyne, "Seeking New Worlds," in Dick and Launius, *Critical Issues*, 8. See Siddiqi, "American Space History," in Dick and Launius, *Critical Issues*, 436; Von Braun, "Introduction," in *Beyond the Solar System*, xiv–xv.

13. On scientism see Stenmark, "What Is Scientism?" 5.

14. Geppert, "Introduction," in *Imagining Outer Space*.

15. Saler, *As If*, 1. See also Gilbert, *Redeeming Culture*, 323.

16. For a similar perspective that questions the allegedly secular values of modernity, see Geppert, "Introduction," in *Imagining Outer Space*. See also Prelinger, *Another Science Fiction*.

17. Launius, "Heroes in a Vacuum"; Neufeld, ed., *Spacefarers*; for further reading see Hersch, *Inventing the American Astronaut*.

18. On transition in representations see Spiller, "Nostalgia for the Right Stuff: Astronauts and Public Anxiety about a Changing Nation," in Neufeld, ed., *Spacefarers*, chapter 3.

Bibliography

ARCHIVAL SOURCES

Bundesarchiv, Potsdam, Germany.

G. Edward Pendray Papers (PP). Mudd Manuscript Library. Princeton University. Correspondence folders.

    Oldenburger, Rufus. "All about the Rockets: Layman's Book Tells Potentialities." *Chicago Sun Book Week*, June 3, 1945, page unknown.

Frederik Pohl Papers (FPP), Syracuse University Library.

Robert A. and Virginia Heinlein Archives. The Heinlein Prize Trust and the University of California, Santa Cruz, Archives, correspondence files.

Wernher von Braun Papers, U.S. Space and Rocket Center, Huntsville, Alabama.

Willy Ley Collection, University of Alabama–Huntsville.

    Interview with Willy Ley, *Night-Beat*, with Mike Wallace, July 1957.

Willy Ley Collection (WLC). Smithsonian National Air and Space Museum, Steven F. Udvar-Hazy Center (NASM Annex, Dulles Airport), Washington, D.C.

    Ley, Willy. Galley Proof, *Inside the Orbit of the Earth*. Box 5, folder 5.

    ———. Manuscript, *Not Beyond Understanding*. Box 7, folder 1.

    "Personal Materials, Moon Book Project." Box 6, folder 9.

    Clippings in WLC, page nos. unknown:

        Ball, Lewis F. "Romance and Common Sense." *Times Dispatch* (Richmond, Va.), September 3, 1955.

        Gannet, Lewis. "Books and Things." *New York Herald Tribune*, June 29, 1951.

        Huxley, Julian. "The Romance of Nature." *Observer*, October 21, 1951.

        Monroe, Burt L. "Pleasant Dip into Scholarship by a Romantic Naturalist." *Louisville Courier-Journal*, August 21, 1955.

Negley, Glenn. "Wily Wilds." *Saturday Review of Literature*, July 26, 1952.

Ray, Ruth D. "Far Away Lands." *Hartford* (Conn.) *Courant*, July 6, 1952.

Unknown author. "How Satellites Will Bring New Communication Era." *Chicago Sun-Times*, December 21, 1958.

————. "Readers Ride to Mythical Destinations." *Fort Wayne* (Ind.) *News-Sentinel*, June 21, 1952.

————. "Rocket Whiz: Ley's 'Trip' Has Been Good." *Long Island Star-Journal*, November 20, 1959.

————. "Scientist Catapulted to Fame." *Athena Press* (Oregon), December 11, 1958.

————. "Turns with a Bookworm." *Tribune Weekly Book Review*, May 20, 1945.

————."Vanguard Career Is Ending; New Missile May Use Name." *Chicago Sun-Times*, October 12, 1958.

————."What German Rocket Experts Did in Russia." *Philadelphia Bulletin*, May 21, 1958.

Walton, Clarence W. "Some Old Secrets of Biology by a Romantic Naturalist." *Virginian Pilot* (Norfolk), August 28, 1955.

SELECTED WORKS BY WILLY LEY

Ley, Willy. "Antlantropa—The Improved Continent." *Modern Science Stories*, February 1939, 99–104.

————. "Anyone Else for Space?" *Galaxy*, June 1964, 110–28.

————. "Any Questions?" *Galaxy*, April 1962, 84–87.

————. "Are We Going to Build a Space Station?" *Galaxy*, December 1962, 125–35.

————. "Atomic Engines for Peace." *Mechanix Illustrated*, March 1946, 44–47, 79.

————. "Atomic Medicine." *Mechanix Illustrated*, April 1946, 41–43, 153–54.

————. "Between Us and Space Travel." *Galaxy*, February 1957, 100–10.

————. *Beyond the Solar System* (with illustrations by Chesley Bonestell). New York: Viking Press, 1964.

————. "The Birth of the Space Station, part 1." *Galaxy*, April 1953, 52–63.

————. "The Birth of the Space Station, part 2." *Galaxy*, May 1953, 84–93.

————. *Bombs and Bombing*. New York: Modern Age, 1941.

————. *The Borders of Mathematics*. New York: Pyramid, 1967.

————. "Botanical Invasion." *Astounding Science Fiction*, February 1940, 91–100.

————. "Calling All Martians!" *Thrilling Wonder Stories*, November 1940, 34–39.

————. "Cold War in Space." *Popular Science*, August 1966, 41–45, 170.

————. "The Conquest of the Deep." *Thrilling Wonder Stories*, June 1938, 36–48.

————. *The Conquest of Space* (with illustrations by Chesley Bonestell). New York: Viking Press, 1949.

————. "The Dawn of the Conquest of Space." *Astounding Stories*, March 1937, 104–10.

———. "Debunking Seversky's 'Victory through Air Power.'" *PM*, December 8, 1942, 2–5.

———. *Das Drachenbuch: Plaudereien von Echsen, Lurchen und Vorweltsauriern.* Leipzig: Thüringer Verlags-Anstalt H. Bartholomäus, 1927.

———. *Die Fahrt ins Weltall.* Leipzig: Hachmeister and Thal, 1926.

———. *Dawn of Zoology.* Engelwood Cliffs, N.J.: Prentice Hall, 1968.

———. *The Days of Creation.* New York: Modern Age, 1941.

———. "Death Under the Sea." *Astounding Science Fiction*, September 1942, 44–53.

———. "Death of the Sun." *Galaxy*, March 1955, 54–65.

———. *The Discovery of the Elements.* New York: Delacorte Press, 1968.

———. "Eight Days in the Story of Rocketry." *Thrilling Wonders Stories*, December 1937, 56–64.

———. "The End of the Rocket Society, part 1." *Astounding Science Fiction*, August 1943, 64–78.

———. "The End of the Rocket Society, part 2." *Astounding Science Fiction*, September 1943, 58–75.

———. *Engineers' Dreams.* New York: Viking, 1951.

———. *Events in Space.* New York: McKay Company, 1969.

———(as Robert Willey). "Fog." *Astounding Science Fiction*, December 1940, 80–102.

———. "For Your Information." *Galaxy*, October 1969, 101–8.

———. "Frau im Mond: Gedanken um Film, Roman und Problem." *Unterhaltungsbeilage*, August 1929, 13–16.

———. "The Future of the Robot Bomb." *Mechanix Illustrated*, November 1944, 142–44.

———. *Gas Planets: The Largest Planets.* New York: McGraw-Hill, 1969.

———. "Geography for Time Travelers." *Astounding Science Fiction*, July 1939, 122–34.

———. *Grudriß einer Geshichte der Rakete.* Leipzig: Hachmeister and Thal, 1932.

———. "How It All Began." *Space World*, June 1961, 23–25, 48–50, 52.

———. "The How of Space Travel." *Galaxy*, October 1955, 60–71.

———. "Improving upon the V-2." *Astounding Science Fiction*, January 1947, 100–20.

———. "Inside the Atom." *Natural History*, October 1945, 350–58.

———. *Inside the Orbit of the Earth.* New York: McGraw-Hill, 1968.

———. "The Kitchen of the Future." *Thrilling Wonder Stories*, October 1939, 96–98.

———. *Konrad Gesner: Leben und Werk.* München: Verlag der Münchner Drucke, 1929.

———. *The Lungfish, the Dodo, and the Unicorn.* New York: Viking Press, 1948.

———. *The Lungfish and the Unicorn: An Excursion into Romantic Zoology.* New York: Modern Age, 1941.

———. "The Man I Didn't Meet." *Galaxy*, August 1961, 131–37.

———. *Man-Made Satellites*. Poughkeepsie, N.Y.: Guild Press, 1957.

———. *Mariner IV to Mars*. New York: New American Library, 1966.

———. *Mars, der Kriegsplanet*. Leipzig: Hachmeister and Thal, 1927.

———. *The Meteorite Craters*. New York: Weybright and Talley, 1968.

———. *Missiles, Moonprobes, and Megaparsecs*. New York: New American Library, 1964.

———. *Die Möglichkeit der Weltraumfahrt: allgemeinverständliche Beiträge zum Raumschiffahrtsproblem*. Leipzig: Hachmeister and Thal, 1928.

———. "Monsters of the Deep." *Galaxy*, February 1959, 94–105.

———. "Moons and Missiles: Space Science Pleas Ignored." *Los Angeles Times*, November 18, 1957, 14.

———. "Moons and Missiles: Von Braun Was Rocket Pioneer." *Los Angeles Times*, November 14, 1957, 2.

———. "The New York World's Fair, 1939." *Life and Letters* 21, no. 20 (July 1939): 50–59.

———. "Nazi's Super Submarine, Too, Is Only Propaganda." *PM*, November 23, 1943, 12–13.

———. "New Weapons Department: Bat Men for Parachute Invasions." *PM*, July 20, 1941, 64.

———. "New Weapons Department: The Rocket-Powered Multiple Cannon for Destroying Enemy Planes." *PM*, June 29, 1941, 38.

———. "The Next Five Years in Space." *Popular Mechanics*, February 1967, 92–97, 212.

———. "Notes on Weapons: V-2, V-1, Me-163." *Technology Review*, January 1945, 169.

———. "The Orbit of Explorer-I." *Galaxy*, October 1968, 93–102.

———. *Our Work in Space*. New York: Macmillan, 1964.

———. "Out of this World by Spaceship." *New York Times*, June 22, 1952, BR1.

———. "Peace or Else!" *Mechanix Illustrated*, February 1946, 44–48.

———(as Robert Willey). "At the Perihelion." *Astounding Science* Fiction, February 1937, 41–66.

———. "Points of View: It's Not Science versus the Humanities." *Instructor*, September 1960, 5.

———. *The Poles*. New York: Time, 1962.

———. "Pseudoscience in Naziland." *Astounding Science Fiction*, May 1947, 90–98.

———. "Push for Pushbutton Warfare." *Astounding Science Fiction*, September 1947, 87–104.

———. *Ranger to the Moon*. New York: New American Library, 1965.

———. "The Reader Speaks." *Wonder Stories*, September 1930, 370.

———. "Review." *Military Affairs* 9 (Winter 1945): 374–75.

———. "The Road to Space Travel, part 1: The Last Twelve Years." *Startling Stories*, March 1949, 68–73.

———. "The Road to Space Travel, part 2: The Next Twelve Years." *Startling Stories*, May 1949, 116–21.

———. "Rocket Artillery." *Astounding Science Fiction*, April 1944, 104–16.

———. "Rockets in Battle." *Technology Review*, December 1946, 95–100.

———. "Rockets into Space." *Science Digest*, January 1957, 65–70.

———. *Rockets: The Future of Travel Beyond the Stratosphere*. New York: Viking, 1944.

———. *Rockets and Space Travel: The Future of Flight Beyond the Stratosphere*. New York: Viking, 1947.

———. *Rockets, Missiles, and Space Travel*. New York: Viking, 1951.

———. *Rockets, Missiles, and Space Travel*. New York: Viking, 1957.

———. *Rockets, Missiles, and Space Travel*. New York: Viking, 1958.

———. *Rockets, Missiles, and Men in Space*. New York: Viking, 1968.

———. *Salamanders and Other Wonders: Still More Adventures of a Romantic Naturalist*. New York: Viking, 1955.

———. "The Satellite Rocket." *Technology Review*, December 1949, 93–95, 112, 114, 116.

———. *Satellites, Rockets, and Outer Space*. New York: New American Library, 1958.

———. "Science and 'Truth.'" *Astounding Science Fiction*, December 1949, 96–111.

———. "Sea of Mystery." *Astounding Science Fiction*, June 1943, 97–108.

———. "The Search for Zero, part 1." *Astounding Science Fiction*, October 1940, 122–34.

———. "The Search for Zero, part 2." *Astounding Science Fiction*, November 1940, 96–108.

———. "See Earth First!" *Startling Stories*, July 1939, 13.

———. *Shells and Shooting*. New York: Modern Age, 1942.

———. "The Shooting Range." *PM*, December 22, 1940.

———. "Some Implications of the Sputniki." *National Review*, December 7, 1957, 515–17.

———. "Soviets Seen Far Ahead in IRBMs." *Los Angeles Times*, December 15, 1957, 15.

———. "The Space Age and Education." *Journal of the New York State School Boards Association*, December 1960, 105.

———. *Space Pilots* (with illustrations by John Polgreen). Poughkeepsie, N.Y.: Guild Press, 1957.

———. *Space Stations*. Adventure in Space series. Poughkeepsie, N.Y.: Guild Press, 1958.

———. *Space Travel*. Adventure in Space series. Poughkeepsie, N.Y.: Guild Press, 1958.

———. "Space Travel by 1960." *Galaxy*, September 1952, 90–99.

———. "Space War." *Astounding Science Fiction*, August 1939, 72–82.

———. "Spaceways: A Fortress in Space." *Space World*, February 1962, 11–14.

———. "Stations in Space." *Amazing Stories*, February 1940, 122–24.

———. "The Story of European Rocketry." *Astronautics* 5, no. 2 (October 1935): 5–9.

———. "A Two-Year Look into Space." *Rotarian*, October 1958, 55–58.

———. "V-2 Rocket Cargo Ship." *Astounding Science Fiction*, May 1945, 99–123.

———. *Visitors from Afar: The Comets*. New York: McGraw-Hill, 1969.

———. "Visitors from the Void." *Astounding Stories*, May 1937, 91–98.

———. "Want a Trip to the Moon?" *Rotarian*, April 1949, 13, 51–54.

———. "War Rockets of the Past." *Coast Artillery Journal*, May–June 1941, 226, 233.

———. "War Weapons: How to Fight Poison Gas." *PM*, June 2, 1942, 5.

———. "War Weapons: Nazi May Use Gas to Test Its Terror Effect." *PM*, May 15, 1942, 9.

———. "War Weapons IX: Rockets." *PM*, January 16, 1942, 20.

———. "War Weapons: The Truth About Germ Warfare." *PM*, March 6, 1942, 9–10.

———. *Watchers of the Skies: An Informal History of Astronomy from Babylon to the Space Age*. New York: Viking, 1963.

———. "What's Wrong with Rockets?" *Amazing Stories*, March 1940, 39, 49–50.

———. "When Will Worlds Collide?" *Galaxy*, July 1952, 75–82.

———. "Who Owns Space?" *Rotarian*, June 1958, 10–13.

———. "The Why of Liquid Propellants for Rockets." *Astronautics* 22 (August–September 1932): 1–2.

———. *Willy Ley's Exotic Zoology*. New York: Viking, 1959.

———. "Willy Ley Recalls 'Captain Future' of Germany." *Science-Fantasy Review* 3, no. 16 (Autumn 1949): 2–4.

———. "The World of October 2052." *Galaxy*, October 1952, 81–93.

———. "You'll Live to See a Spaceship: The First Rocket Will Leave the Earth in 1965 or 1970." *Today: The Philadelphia Inquirer Magazine*, November 16, 1951, 9–11.

Ley, Willy, with Wolfgang Both and Klaus Scheffler. *Die Starfield Company*. Berlin: Shayol, 2011.

Ley, Willy, and P. E. Cleator. "The Rocket Controversy." *Armchair Science*, April 1935, 17–19.

Ley, Willy, and Wernher von Braun. *The Conquest of Space: A Conversation between Wernher von Braun and Willy Ley* (LP recording). New York: Vox, 1959. Accessible at archive.org.

———. *The Exploration of Mars*. New York: Viking, 1956.

Ley, Willy, ed. *Harnessing Space*. New York: Macmillan and Collier-Macmillan, 1963.

PUBLISHED SOURCES

Adas, Michael. *Machines as the Measure of Men: Science, Technology, and the Ideologies of Western Dominance.* Ithaca: Cornell University Press, 1989.

Agassi, Joseph. *Towards a Historiography of Science.* The Hague: Mouton and Company, 1963.

Alpers, Benjamin L. *Dictators, Democracy, and American Public Culture: Envisioning the Totalitarian Enemy, 1920s–1950s.* Chapel Hill: University of North Carolina Press, 2003.

Amrine, Michael. "Those Cities of Gold." *New York Times,* June 22, 1952, BR6.

Andrews, James T., and Asif A. Siddiqi, eds. *Into the Cosmos: Space Exploration and Soviet Culture.* Pittsburgh: University of Pittsburgh Press, 2011.

———. *Science for the Masses: The Bolshevik State, Public Science, and the Popular Imagination in Soviet Russia, 1917–1934.* College Station: Texas A&M University Press, 2003.

Anonymous. "Ley, Willy." In *Current Biography.* New York: H. W. Wilson, 1941, 512–13.

———. "Ley, Willy." In *The Scanner: A Monthly Publication of the Monmouth County (N.J.) Sub-Section Institute of Radio Engineers),* February 1956, 3–5.

———. "Ley, Willy." In *Twentieth Century Authors First Supplement.* New York: H. W. Wilson, 1955, 580–81.

———. "Ley, Willy." In *World Authors: 1900–1950,* rev. ed., 4 vols., ed. Martin Seymour-Smith and Andrew C. Caimans. New York: H. W. Wilson, 1996.

Anton, Ludwig. *Brücken über den Weltenraum.* Bad Rothefelde: Holzwarth-Verlag, 1922.

Asimov, Isaac. *A Short History of Biology.* Garden City, N.Y.: Natural History Press, 1964.

———. *Asimov's Biographical Encyclopedia of Science and Technology.* Garden City, N.Y.: Doubleday, 1965.

———. *In Joy Still Felt: The Autobiography of Isaac Asimov.* Garden City, N.Y.: Doubleday, 1980.

———. *In Memory Yet Green: The Autobiography of Isaac Asimov, 1920–1954.* Garden City, N.Y.: Doubleday, 1979.

Barksdale Clowes, Barbara. *Brainpower for the Cold War: The Sputnik Crisis and National Defense Act of 1958.* Westport, Conn.: Greenwood Press, 1981.

Audoin-Rouzeau, Stéphane, and Annette Becker. *Understanding the Great War.* New York: Hill and Wang, 2000.

Baughman, James L. *The Republic of Mass Culture: Journalism, Filmmaking, and Broadcasting in America since 1941.* Baltimore: Johns Hopkins University Press, 1992.

Bell, Philip, Bruce Lewenstein, Andrew W. Shouse, and Michael A. Feder, eds. *Learning Science in Informal Environments: People, Places, and Pursuits.* Washington, D.C.: National Academies Press, 2009.

Benjamin, Marina. *Rocket Dreams: How the Space Age Shaped Our Vision of a World Beyond*. New York: Free Press, 2003.

Biddle, Wayne. *The Dark Side of the Moon: Wernher von Braun, the Third Reich, and the Space Race*. New York: W. W. Norton, 2009.

Blakesley, Richard. "We're Off to the Moon! A Travelog." *Chicago Daily Tribune*, October 16, 1949, H4.

Bölsche, Wilhelm. *Love-Life in Nature: The Story of the Evolution of Love*, trans. Cyril Brown. New York: Albert and Charles Boni, 1926.

Bonestell, Chesley. "Tribute to Willy Ley." *Spaceflight* 11, November 1969, 408–9.

Botsford, Harry. "The Rockets Rain Down." *Mechanix Illustrated*, June 1945, 47–49.

Bowler, Peter. *Science for All: The Popularization of Science in Early Twentieth-Century Britain*. Chicago: University of Chicago Press, 2009.

Boyer, Paul. *By the Bomb's Early Light: American Thought and Culture at the Dawn of the Atomic Age*. Chapel Hill: University of North Carolina Press, 1985.

Breit, Harvey. "Talk with Willy Ley." *New York Times*, July 22, 1951, 156.

Bronowski, Jacob. *Science and Human Values*. New York: J. Messner, 1956.

Brown, Theodore M. "Review." *American Scientist* 58 (May–June 1970): 335–36.

Bulkeley, Rip. *The Sputniks Crisis and Early United States Space Policy: A Critique of Historiography*. Bloomington: Indiana University Press, 1991.

Bullock, Marcus P. *The Violent Eye: Ernst Jünger's Visions and Revisions on the European Right*. Detroit: Wayne State University Press, 1992.

Burnham, John C. *How Superstition Won and Science Lost: Popularizing Science and Health in the United States*. New Brunswick, N.J.: Rutgers University Press, 1987.

Burns, Eric. *Invasion of the Mind Snatchers: Television's Conquest of America in the Fifties*. Philadelphia: Temple University Press, 2010.

Bush, Vannevar. *Modern Arms and Free Men*. New York: Simon and Schuster, 1949.

———. "Science, Strength, and Stability." *Technology Review*, July 1946, 553–56, 586, 588.

———. "The Scientific Way." *Technology Review*, June 1947, 463–64, 482, 484, 486.

Butsch, Richard. *The Citizen Audience: Crowds, Publics, and Individuals*. London: Routledge, 2008.

———. *The Making of American Audiences from Stage to Television, 1750–1900*. Cambridge: Cambridge University Press, 1990.

Cagle, Fred R. "Review." *Copeia*, February 29, 1956, 71.

Campbell, John W. *John W. Campbell Letters*, vols. 1 and 2. Franklin, Tenn.: Ac Projects, 1985.

Cannon, Walter F. "Review." *Isis* 56, no. 3 (Autumn 1965): 391–92.

Carlson, Robert E. "Review." *Isis* 54, no. 1 (March 1963): 168–69.

Chambers, David Wade. "History of Science on the Silver Screen." *Isis* 57, no. 4 (Winter 1966): 494–97.

Cheng, John. *Astounding Wonder: Imagining Science and Science Fiction in Interwar America*. Philadelphia: University of Pennsylvania Press, 2012.

Chesneaux, Jean. "Jules Verne's Image of the United States." *Yale French Studies* 43 (1969): 111–27.

Christman, Ruth G., ed. *Soviet Science*. Washington, D.C.: American Association for the Advancement of Science, 1952.

Clarke, Arthur C. *The Exploration of Space*. New York: Harper, 1951.

Cleator, P. E. "Tribute to Willy Ley." *Spaceflight* 11, November 1969, 408.

Cohen, I. Bernard. "Guide for Space Travel." *New York Times*, December 1, 1963, 496.

———. "Review." *Isis* 56, no. 2 (Summer 1965): 240–42.

Collins, [no initial]. "What Happened to the Dodo Bird and Where Did the Waldrapp Go?" *Houston Chronicle*, August 14, 1955.

Compton, Arthur H. "Education for Peace." *Technology Review*, July 1948, 501–3, 530.

Conant, James B. *On Understanding Science: An Historical Approach*. New Haven, Conn.: Yale University Press, 1947.

Cornell, Bernice. "More Information, Please!" *Writers' Markets and Methods*, February 1942, 7–8, 14–15.

Cornwell, John. *Hitler's Scientists: Science, War, and the Devil's Pact*. New York: Penguin Books, 2004.

Crouch, Tom. *Aiming for the Stars: The Dreamers and Doers of the Space Age*. Washington D.C.: Smithsonian Institution Press, 1999.

Daum, Andreas. "Science, Politics, and Religion: Humboldtian Thinking and the Transformation of Civil Society in Germany, 1830–1870." *Osiris* 17 (2002): 107–40.

———. "Varieties of Popular Science and the Transformations of Public Knowledge: Some Historical Reflections." *Isis* 100 (2009): 319–32.

———. *Wissenschaftspopularisierung im 19. Jahrhundert: Bürgerliche Kultur, naturwissenschaftliche Bildung und die deutsche Öffentlichkeit, 1848–1914*. Munich: Oldenbourg, 2002.

Davenport, Basil. "Watchers of the Skies." *Book of the Month Club News*, November 1963, 8.

Davis, Belinda J. *Home Fires Burning: Food, Politics, and Everyday Life in World War I Berlin*. Chapel Hill: University of North Carolina Press, 2000.

Davis, Susan G. *Spectacular Nature: Corporate Culture and the Sea World Experience*. Berkeley: University of California Press, 1997.

De Camp, L. Sprague. *Time and Chance: An Autobiography*. Hampton Falls, N.H.: Donald M. Grant, Publisher, 1996.

De Camp, L. Sprague, and Willy Ley. *Lands Beyond*. New York: Rinehart, 1952.

Del Rey, Lester. "Credo: The First Citizen of the Moon." *Galaxy*, September 1969, 151–57.

Derleth, August. "A Scientific Adventure in Romance." *Chicago Daily Tribune*, January 28, 1951, H4.

———. "The Romantic Naturalist." *Chicago Daily Tribune*, August 28, 1955, B3.

Dick, Steven J., ed. *NASA's First Fifty Years: Historical Perspectives.* Washington, D.C.: National Aeronautics and Space Administration, 2006.

———, ed. *Remembering the Space Age: Proceedings of the 50th Anniversary Conference.* Washington, D.C.: National Aeronautics and Space Administration, 2008.

Dick, Steven J., and Roger D. Launius, eds. *Critical Issues in the History of Spaceflight.* Washington, D.C.: National Aeronautics and Space Administration, 2006.

Dickson, Paul. *Sputnik: The Shock of the Century.* New York: Walker and Company, 2001.

"Display Ad 116." *New York Times*, May 21, 1944, BR9.

"Display Ad 191." *Chicago Daily Tribune*, July 28, 1957, G3.

"Display Ad 220." *New York Times*, May 25, 1947, BR31.

"Display Ad 330." *New York Times*, July 28, 1968, 262.

"Display Ad 378." *New York Times*, May 12, 1968, BR24.

"Display Ad 971." *New York Times*, September 30, 1951, 178.

Divine, Robert. *The Sputnik Challenge: Eisenhower's Response to the Soviet Satellite.* Oxford: Oxford University Press, 1993.

Doel, Ronald E., and Thomas Söderqvist, eds. *The Historiography of Contemporary Science, Technology, and Medicine: Writing Recent Science.* London: Routledge, 2006.

Earley, George W. "Beyond the Earth." *Hartford Courant*, June 16, 1968, BBR.

Emme, Eugene M. "Space, Past, Future." *Science* 161 (August 30, 1968): 874–75.

Essers, I. *Max Valier: A Pioneer of Space Travel.* Washington, D.C.: National Aeronautics and Space Administration, 1976.

———. *Ein Vorkämpfer der Weltraumfahrt, 1895–1930.* Düsseldorf: VDI-Verlag, 1968.

Face the Nation, Volume 1, 1954–1955: The Collected Transcripts from the CBS Radio and Television Broadcasts. New York: Columbia Broadcasting System, 1972.

Fink, John. "Space Expert Helps Kids to Learn about It." *Chicago Daily Tribune*, August 25, 1957, SW14.

Fischer, William B. *The Empire Strikes Out: Kurd Lasswitz, Hans Dominik, and the Development of German Science Fiction.* Bowling Green, Ohio: Bowling Green State University Press, 1986.

Fisher, Peter S. *Fantasy and Politics: Visions of the Future in the Weimar Republic.* Madison: University of Wisconsin Press, 1991.

Forman, Paul. "Scientific Internationalism and the Weimar Physicists: The Ideology and Its Manipulation in Germany after World War 1." *Isis* 64, no. 2 (June 1973): 150–80.

Frayling, Christopher. *Mad, Bad, and Dangerous? The Scientist and the Cinema.* New York: Reaktion Books, 2006.

Freeman, Marsha. *How We Got to the Moon: The Story of the German Space Pioneers.* Washington, D.C.: 21st Century Science Associates, 1993.

Friedrich, Otto. *Before the Deluge: A Portrait of Berlin in the 1920s.* New York: Harper Perennial, 1972.

Fritzsche, Peter. *A Nation of Fliers: German Aviation and the Popular Imagination.* Cambridge: Harvard University Press, 1992.

———. *Reading Berlin, 1900.* Cambridge: Harvard University Press, 1996.

Gail, Otto Willi. *Der Schuss ins All: Ein Roman von Morgan.* Breslau: Bergstadtverlag, 1925.

Gardner, Martin. "Bridge to the Moon." *Scientific Monthly* 70 (January 1950): 71–72.

———. "Through Pathless Realms." *Scientific Monthly* 66 (January 1948): 78.

Gardner, Thomas S. "Review." *Scientific Monthly* 82, no. 3 (March 1956): 143.

Gatta, John. *Making Nature Sacred: Literature, Religion, and Environment in America from the Puritans to the Present.* Oxford: Oxford University Press, 2004.

Gay, Peter. *Weimar Culture: The Outsider as Insider.* New York: W. W. Norton, 1968.

Geppert, Alexander C. T., ed. *Imagining Outer Space: European Astroculture in the Twentieth Century.* New York: Palgrave Macmillan, 2012.

Gibbons, Roy. "All Aboard for Trip to Moon or Some Such." *Chicago Daily Tribune,* August 12, 1951, B10.

Gibson, James William. *A Reenchanted World: The Quest for a New Kinship with Nature.* New York: Henry Holt and Company, 2009.

Gilbert, James. *Redeeming Culture: American Religion in an Age of Science.* Chicago: University of Chicago Press, 1997.

Gilman, William. *Science: U.S.A.* New York: Viking Press, 1965.

Gingerich, Owen. "Watchers of the Skies." *Sky and Telescope,* September 1964, 164.

Goddard, Calvin. "Review." *Military Affairs* 6 (Spring 1942): 50–51.

Goddard, Robert H. *A Method of Reaching Extreme Altitudes.* Washington, D.C.: Smithsonian Institution, 1919.

Golinski, Jan. *Making Natural Knowledge: Constructivism and the History of Science.* Chicago: University of Chicago Press, 2005.

Gordin, Michael. *The Pseudoscience Wars: Immanuel Velikovsky and the Birth of the Modern Fringe.* Chicago: University of Chicago Press, 2012.

Hahn, Otto. *Otto Hahn: A Scientific Autobiography.* Trans. Willy Ley. New York: Charles Scribner's Sons, 1966.

Hall, A. Rupert. *From Galileo to Newton, 1630–1720.* New York: Harper and Row, 1963.

———. *The Scientific Revolution.* Boston: Beacon Press, 1954.

Hall, A. Rupert, and Marie Boas Hall. *A Brief History of Science*. New York: New American Library, 1964.

Hall, Marie Boas. *The Scientific Renaissance, 1450–1630*. New York: Harper and Brothers, 1962.

Halliwell, Martin. *Romantic Science and the Experience of Self: Transatlantic Cross-currents from William James to Oliver Cromwell*. Studies in European Cultural Transition, vol. 2. Aldershot, U.K.: Ashgate, 1999.

Harrington, Anne. *Reenchanted Science: Holism in German Culture from Wilhelm II to Hitler*. Princeton: Princeton University Press, 1996.

Haslam, Robert T. "World Energy and World Peace." *Technology Review*, July 1948, 493–96, 510.

Haynes, Roslynn D. *From Faust to Strangelove: Representations of the Scientist in Western Literature*. Baltimore: Johns Hopkins University Press, 1994.

Heinlein, Robert. "Baedeker of the Solar System." *Saturday Review of Literature*, December 24, 1949, 9–10.

Hellman, C. Doris. "Review." *Isis* 55, no. 3 (September 1964): 376–77.

Herf, Jeffrey. *Reactionary Modernism: Technology, Culture, and Politics in Weimar and the Third Reich*. Cambridge: Cambridge University Press, 1986.

Hersch, Matthew H. *Inventing the American Astronaut*. New York: Palgrave Macmillan, 2012.

Hollinger, David A. "Science as a Weapon in Kulturkämpfe in the United States During and After World War II." *Isis* 86, no. 3 (September 1995): 440–54.

———. *Science, Jews, and Secular Culture: Studies in Mid-Twentieth Century American Intellectual History*. Princeton: Princeton University Press, 1996.

———. "The Unity of Knowledge and the Diversity of Knowers: Science as an Agent of Cultural Integration Between the Two World Wars." *Pacific Historical Review* 80 (May 2011): 211–30.

Holmes, Frederic L. "Review." *Isis* 56, no. 2 (Summer 1965): 228–29.

Holmes, Richard. *The Age of Wonder: The Romantic Generation and the Discovery of the Beauty and Terror of Science*. New York: Pantheon, 2008.

Hopwood, Nick. "Producing a Socialist Popular Science in the Weimar Republic." *History Workshop Journal* 41 (1996): 117–53.

Hoyle, Fred. *Of Men and Galaxies*. Seattle: University of Washington Press, 1964.

Iwanska, Alicja. "Popular Science and Race." *Phylon* 14 (1953): 98.

Jackson, Joseph Henry. "Bookman's Notebook." *Los Angeles Times*, February 1, 1951, A5.

Jewett, Andrew. *Science, Democracy, and the American University: From the Civil War to the Cold War*. Cambridge: Cambridge University Press, 2013.

Joravsky, David. "The History of Technology in Soviet Russia and Marxist Doctrine." *Technology and Culture* 2 (Winter 1961): 5–10.

———. "Review." *Isis* 51, no. 3 (September 1960): 348–53.

————. *Soviet Marxism and Natural Science, 1917–1932*. New York: Columbia University Press, 1961.

————. "Soviet Views on the History of Science." *Isis* 46, no. 1 (March 1955): 3–13.

Judd, Richard. *The Untilled Garden: Natural History and the Spirit of Conservation in America, 1740–1840*. New York: Cambridge University Press, 2009.

Kaempffert, Waldemar. "Ranging Beyond the Stratosphere." *New York Times*, May 21, 1944, BR4.

————. "Science in the Totalitarian State." *Foreign Affairs* 19 (January 1941): 433–41.

Kater, Michael H. "The Work Student: A Socio-Economic Phenomenon of Early Weimar Germany." *Journal of Contemporary History* 10 (1975): 70–94.

Keenan, Joseph H. "Education for Freedom." *Technology Review*, February 1951, 195, 220, 222.

Kelly, Alfred. *The Descent of Darwin: The Popularization of Darwinism in Germany, 1860–1914*. Chapel Hill: University of North Carolina Press, 1981.

Kilgore, De Witt Douglas. *Astrofuturism: Science, Race, and Visions of Utopia in Space*. Philadelphia: University of Pennsylvania Press, 2003.

Killen, Andreas. *Berlin Electropolis: Shock, Nerves, and German Modernity*. Berkeley: University of California Press, 2006.

Killian, James R. Jr. "Academic Freedom and Communism." *Technology Review*, May 1949, 432.

————. "Our Shared Convictions." *Technology Review*, July 1950, 503–5, 550.

Kirby, David K. *Lab Coats in Hollywood: Science, Scientists, and Cinema*. Cambridge, Mass: MIT Press, 2011.

————. "Science Consultants, Fictional Films, and Scientific Practice." *Social Studies of Science* 33 (2002): 231–68.

Koestler, Arthur. *The Sleepwalkers: A History of Man's Changing Vision of the Universe*. New York: Hutchinson, 1959.

Kracauer, Siegfried. "Lokomotive über der Friedrichstraße." *Schriften*, vol. v.iii, ed. Inka Mülder-Bach. Frankfurt am Main: Suhrkamp, 1990.

Krutch, Joseph Wood. "Naturally It's Strange." *New York Times*, August 29, 1955, BR6.

————. "Responsible Fantasy." *Nation*, October 1, 1949, 330.

————. "Review." *Nation*, August 25, 1951, 155.

Kuhn, Thomas S. *The Copernican Revolution: Planetary Astronomy in the Development of Western Thought*. Cambridge: Harvard University Press, 1957.

————. *The Structure of Scientific Revolutions* (1962), 4th edition. Chicago: University of Chicago Press, 2012.

LaFollette, Marcel C. *Making Science Our Own: Public Images of Science, 1910–1955*. Chicago: University of Chicago Press, 1990.

————. *Science on American Television: A History*. Chicago: University of Chicago Press, 2013.

———. *Science on the Air: Popularizers and Personalities on Radio and Early Television*. Chicago: University of Chicago Press, 2008.

———. "A Survey of Science Content in U.S. Television Broadcasting, 1940s through 1950s: The Exploratory Years." *Science Communication* 24 (September 2002): 34–71.

Lang, Fritz. "Sci-Fi Film-maker's Debt to Rocket Man Willy Ley." *Los Angeles Times*, July 27, 1969, P24.

Lasswitz, Kurd. *Auf Zwei Planeten*. Weimar: Felber, 1897.

Launius, Roger D. *Frontiers of Space Exploration*. Westport, Conn.: Greenwood Press, 1998.

———. "Heroes in a Vacuum: The Apollo Astronaut as Cultural Icon." *Florida Historical Quarterly* 87 (Fall 2008): 147–209.

———. "The Historical Dimension of Space Exploration: Reflections and Possibilities." *Space Policy* 16 (2000): 23–38.

Launius, Roger D., John M. Logsdon, and Robert W. Smith, eds. *Reconsidering Sputnik: Forty Years since the Soviet Satellite*. Amsterdam: Harwood, 2000.

Launius, Roger D., and Howard E. McCurdy, eds. *Spaceflight and the Myth of Presidential Leadership*. Urbana: University of Illinois Press, 1997.

Leane, Elizabeth. *Reading Popular Physics: Disciplinary Skirmishes and Textual Strategies*. Burlington, Vt.: Ashgate, 2007.

Leed, Eric. *No Man's Land: Combat Identity and World War I*. Cambridge: Cambridge University Press, 1979.

Leiber, Fritz R. Jr. "Some Strange Animals—Real And Faked." *Chicago Daily Tribune*, October 31, 1948, E9.

Lieber, Leslie. "Whispering Willy." *Los Angeles Times*, June 26, 1955, H37.

Lightman, Bernard. *Victorian Popularizers of Science: Designing Nature for New Audiences*. Chicago: University of Chicago Press, 2007.

———. "'The Voices of Nature': Popularizing Victorian Science." In *Victorian Science in Context*, ed. Bernard Lightman, 187–211. Chicago: University of Chicago Press, 1997.

Lindeman, Eduard C. "Introduction." In *The Scientific Spirit and Democratic Faith*, ed. Eduard C. Lindeman, ix–xii. New York: King's Crown Press, 1944.

Logsdon, John M., Linda J. Lear, Jannelle Warren Findley, Ray A. Williamson, and Dwayne A. Day, eds. *Exploring the Unknown: Selected Documents in the History of the U.S. Civil Space Program*, vol. 1: *Organizing for Exploration*. Washington, D.C.: National Aeronautics and Space Administration, 1995.

Lundberg, George A. *Can Science Save Us?* New York: Longmans, Green and Company, 1947.

Maddock, A. G. "Review." *Isis* 58, no. 4 (Winter 1967): 583–84.

Mann, W. H. "Fancy and Fact." *Scientific Monthly* 68 (May 1949): 358–59.

Marché, Jordan C. *Theaters of Time and Space: American Planetaria, 1930–1970*. New Brunswick, N.J.: Rutgers University Press, 2005.

Markley, Robert. *Dying Planet: Mars in Science and the Imagination*. Durham: Duke University Press, 2005.

Marks, Lionel S. "Jet Propulsion Theories." *Scientific Monthly* 61 (September 1945): 240–42.

Mauro, James. *Twilight at the World of Tomorrow: Genius, Madness, Murder and the 1939 World's Fair on the Brink of War*. New York: Ballantine Books, 2010.

May, Mark A. "The Morale Code of the Scientist." In *The Scientific Spirit and Democratic Faith*, ed. Eduard C. Lindeman, 40–46. New York: King's Crown Press, 1944.

Mayerhöfer, Josef. "Review." *Isis* 56, no. 3 (Autumn 1965): 368–69.

McCray, W. Patrick. *Keep Watching the Skies! The Story of Operation Moonwatch and the Dawn of the Space Age*. Princeton: Princeton University Press, 2008.

———. *The Visioneers: How a Group of Elite Scientists Pursued Space Colonies, Nanotechnologies, and a Limitless Future*. Princeton: Princeton University Press, 2013.

McCurdy, Howard E. *Space and the American Imagination*. Washington, D.C.: Smithsonian Institution Press, 1997.

McDougall, Walter A. *The Heavens and the Earth: A Political History of the Space Age*. 1985; Baltimore: Johns Hopkins University Press, 1997.

McNash, C. W. "Proposed Altitude Rocket Hops." *Popular Aviation*, October 1936, 7.

McQuaid, Kim. "Sputnik Reconsidered: Image and Reality in the Early Space Age." *Canadian Review of American Studies* 37 (2007): 371–401.

Merchant, Carolyn. *The Death of Nature: Women, Ecology, and the Scientific Revolution*. New York: Harper and Row, 1980.

Middleton, Mary. "Women Plan May Benefit for Hospital." *Chicago Daily Tribune*, March 25, 1958, A4.

Mieczkowski, Yanek. *Eisenhower's Sputnik Moment: The Race for Space and World Prestige*. Ithaca: Cornell University Press, 2013.

Miles, Wyndham. "Review." *Isis* 41, no. 2 (July 1950): 237–38.

Miller, Ron, and Frederick C. Durant III (with Melvin Schuetz). *The Art of Chesley Bonestell*. New York: Collins and Brown, 2001.

Mills, Deborah J. "Review." *Isis* 56, no. 3 (Autumn 1965), 373.

Montagu, M. F. Ashley. *Man's Most Dangerous Myth: The Fallacy of Race*. New York: Columbia University Press, 1942.

———. "Review of *Adventures of a Biologist* by J.B.S. Haldane." *Isis* 33, no. 2 (June 1941): 298–99.

Moskowitz, Sam. "Willy Ley: Forgotten Prophet of the Space Age." *Fantasy Review* 99 (March 1987): 12–15, 52–53.

———. "Willy Ley in the U.S.A, part III: Losing the Last One." *Fantasy Review* 102 (June 1987): 17–20.

———. "The Willy Ley Story." *Worlds of Tomorrow*, May 1966, 30–42.

Mosse, George. "Review." *Isis* 51, no. 3 (September 1960): 361–62.

Nagl, Manfred. "National Peculiarities of German SF." *Science Fiction Studies* 8, no. 1 (1981): 29–34.

———. "SF, Occult Sciences, and Nazi Myths." *Science Fiction Studies* 3, no. 3 (1974): 185–97.

Nebel, Rudolf. *Die Narren von Tegel: Ein Pioneer der Raumfahrt.* Düsseldorf: Droste, 1972.

Neufeld, Michael J. "Creating a Memory of the German Rocket Program for the Cold War." In *Remembering the Space Age*, ed. Steven J. Dick, 71–88. Washington, D.C.: National Aeronautics and Space Administration, 2008.

———. "The Excluded: Hermann Oberth and Rudolf Nebel in the Third Reich." In *History of Rocketry and Astronautics: Proceedings of the Twenty-Eighth and Twenty-Ninth History Symposia of the International Academy of Astronautics*, ed. Donald C. Elder and Christophe Rothmund, 209–22. San Diego: AAS, 2001.

———. "The Reichswehr, the Rocket, and the Versailles Treaty: A Popular Myth Reexamined." *Journal of the British Interplanetary Society* 53 (May–June 2000): 163–72.

———. *The Rocket and the Reich: Peenemünde and the Coming of the Ballistic Missile Era.* Cambridge: Harvard University Press, 1995.

———. "Spaceflight Advocacy from Weimar to Disney." In *1998 National Aerospace Conference Proceedings.* Dayton, Ohio: Wright State University, 1999.

———. *Von Braun: Dream of Space, Engineer of War.* New York: Vintage Books, 2007.

———. "Weimar Culture and Futuristic Technology: The Rocketry and Spaceflight Fad in Germany, 1923–1933." *Technology and Culture* 31, no. 4 (October 1990): 725–52.

———. "Wernher von Braun, the SS, and Concentration Camp Labor: Questions of Moral, Political, and Criminal Responsibility." *German Studies Review* 25, no. 1 (February 2002): 58–78.

Neufeld, Michael J., ed. *Spacefarers: Images of Astronauts and Cosmonauts in the Heroic Age of Spaceflight.* Washington, D.C.: Smithsonian Institution Press, 2013.

Nyart, Lynn K. *Modern Nature: The Rise of the Biological Perspective in Germany.* Chicago: University of Chicago Press, 2009.

Nye, David. *American Technological Sublime.* Cambridge, Mass.: MIT Press, 1994.

O'Neill, John J. "Writer Charges U.S. with Curb on Science." *New York Times*, August 14, 1941, 15.

Oberth, Hermann. *Die Rakete zu den Planeten räumen.* München: R. Oldenbourg, 1923.

———. *Wege zur Raumschiffahrt.* München: R. Oldenbourg, 1929.

Olby, R. C., G. N. Cantor, J.R.R. Christie, and M.J.S. Hodge, eds. *Companion to the History of Modern Science.* London: Routledge, 1990.

Oppenheimer, J. Robert. "Physics in the Contemporary World." *Technology Review*, February 1948, 201–4, 231–34, 238.

Pandora, Katherine. "Popular Science in National and Transnational Perspective: Suggestions from the American Context." *Isis* 100, no. 2 (June 2009): 346–58.

———. *Rebels within the Ranks: Psychologists' Critique of Scientific Authority and Democratic Realities in New Deal America*. Cambridge: Cambridge University Press, 1997.

Pandora, Katherine, and Karen A. Rader. "Science in the Everyday World: Why Perspectives from the History of Science Matter." *Isis* 99, no. 2 (June 2008): 350–64.

Parrish, Susan Scott. *American Curiosities: Cultures of Natural History in the Colonial British Atlantic World*. Chapel Hill: University of North Carolina Press, 2006.

Parry, Albert. *Russia's Rockets and Missiles*. London: Macmillan, 1960.

Patterson, William H. *Robert A. Heinlein: In Dialogue with His Century*. New York: Tor Books, 2011.

Pendray, G. Edward. "Age of Rocket Power." *Coronet*, March 1946, 149–61.

———. *The Coming Age of Rocket Power*. New York: Harper and Brothers, 1945.

———. "Next Stop the Moon." *Collier's*, September 1946; repr. *Coast Artillery Journal*, January–February 1947, 48–51.

———. "Number One Rocket Man." *Scientific American*, May 1938.

———. "The Persistent Man." *Coast Artillery Journal* 91 (January–February 1948): 52–56.

———. "The Reaction Engine." *Popular Science*, May 1945, 70–72.

———. "Robert H. Goddard." *Science*, November 23, 1945, 521–22.

———. "Skyrockets Grow Up." *This Week*, October 10, 1943, 4–6.

———. "To the Moon via Rocket?" *Sky*, November 1936.

Pettitt, Edison. "Review." *Publications of the Astronomical Society of the Pacific* 61, no. 363 (December 1949): 269–71.

Pfeiffer, John E. "Round-Trip Ticket to the Moon." *New York Times*, September 25, 1949, BR29.

———. "Traveling in Space." *New York Times*, June 24, 1956, BR6.

Phillips, H. B. "What Is Democracy?" *Technology Review*, June 1948, 433–34, 456, 458, 460.

Poore, Charles. "Books of the Times." *New York Times*, June 19, 1952, 25.

Prelinger, Megan. *Another Science Fiction: Advertising the Space Race, 1957–1962*. New York: Blast Books, 2010.

Prescott, Orville. "Books of the Times." *New York Times*, December 29, 1948, 19.

———. "Books of the Times." *New York Times*, December 26, 1949, 27.

———. "Books of the Times." *New York Times*, February 12, 1951, 20.

Reingold, Nathan. "Review." *Isis* 57, no. 2 (Summer 1966): 275–76.

Reisch, George A. *How the Cold War Transformed the Philosophy of Science: To the Icy Slopes of Logic*. Cambridge: Cambridge University Press, 2005.

Richards, Robert J. *The Romantic Conception of Life: Science and Philosophy in the Age of Goethe.* Chicago: University of Chicago Press, 2002.

Richardson, Robert S. "Review." *Publications of the Astronomical Society of the Pacific* 56, no. 330 (June 1944): 134.

Rudolph, John. *Scientists in the Classroom: The Cold War Reconstruction of American Science Education.* New York: Palgrave, 2002.

Ryan, Cornelius, ed. *Across the Space Frontier.* New York: Viking, 1952.

———. *Conquest of the Moon.* New York: Viking, 1953.

Rydell, Robert W., and Laura B. Schiavo, eds. *Designing Tomorrow: America's World's Fairs of the 1930s.* New Haven: Yale University Press, 2010.

Sachs, Aaron. *The Humboldt Current: Nineteenth-Century Exploration and the Roots of American Environmentalism.* New York: Viking, 2006.

Saler, Michael. *As If: Modern Enchantment and the Literary Prehistory of Virtual Reality.* Oxford: Oxford University Press, 2012.

———. "Modernity and Enchantment: A Historiographic Review." *American Historical Review* 111 (2006): 692–716.

Santillana, Giorgio de, and Stillman Drake. "Arthur Koestler and His Sleepwalkers." *Isis* 50, no. 3 (September 1959): 255–60.

Saunders, Thomas J. *Hollywood in Berlin: American Cinema and Weimar Germany.* Berkeley: University of California Press, 1994.

Sarton, George. "Preface to Volume XXXIII: To the Republic of Letters." *Isis* 33, no. 1 (March 1941): 3–4.

Sax, Karl. "Review: Conway Zirkle, *The Death of a Science in Russia.*" *Isis* 41, no. 2 (July 1950): 238–39.

Scheuer, Philip K. "Hollywood Will Reach Moon First." *Los Angeles Times,* December 11, 1949, E1–E2.

Schwartz, Harry. "Books of the Times." *New York Times,* December 6, 1963, 32.

Secord, James A. "Knowledge in Transit." *Isis* 95, no. 4 (December 2004): 654–72.

Seversky, Alexander P. de. *Victory through Air Power.* New York: Simon and Schuster, 1942.

Shapiro, H. L. "Man and the Atom." *Natural History,* October 1945, 350.

Sharpe, Mitchell R., and Frederick I. Ordway III. *The Rocket Team.* Boston: MIT Press, 1982.

Sherman, Gene. "Willy Ley Appalled by U.S. Rocket Lag." *Los Angeles Times,* May 4, 1959, B5.

Sherwood, Taylor S. *The Alchemists: Founders of Modern Chemistry.* New York: Henry Schuman, 1949.

Sherwood, Thomas S. "Science in Education." *Technology Review,* December 1948, 98–100.

Shteir, Ann B. *Cultivating Women, Cultivating Science: Flora's Daughters and Botany in England, 1760 to 1860.* Baltimore: Johns Hopkins University Press, 1996.

Siddiqi, Asif S. "American Space History: Legacies, Questions, and Opportunities for Future Research." In *Critical Issues in the History of Spaceflight*, 434. ed. Steven J. Dick and Roger D. Launius. Washington, D.C.: National Aeronautics and Space Administration, 2006.

———. *The Red Rockets' Glare: Spaceflight and the Soviet Imagination*. Cambridge: Cambridge University Press, 2010.

Smith, David R. "They're Following Our Script: Walt Disney's Trip to Tomorrowland." *Future*, May 1978, 54–62, 54.

Smith, Terry. *Making the Modern: Industry, Art, and Design in America*. Chicago: University of Chicago Press, 1993.

Snow, Charles Percy. *The Two Cultures and the Scientific Revolution*. Rede Lecture, 1959. Cambridge: Cambridge University Press, 1959.

Solomon, Stan. "Oldies and Oddities." *Air & Space*, December 1993–January 1994, 87.

Spohr, Carl W. "The Final War." *Wonder Stories*, March and April 1932.

Stannard, Jerry. "Review." *Isis* 59, no. 3 (Autumn 1968): 334–35.

———. "Review." *Science* 163:3862 (January 3, 1969): 64–65.

Starr, Paul. *The Creation of the Media: Political Origins of Modern Communications*. New York: Basic Books, 2004.

Starrett, Vincent. "Books Alive." *Chicago Daily Tribune*, November 21, 1948, G4.

Stenmark, Mikael. "What Is Scientism?" *Religious Studies* 33 (March 1997): 15-32.

Strodder, Chris. *The Disney Encyclopedia: The Unofficial, Unauthorized, and Unprecedented History of Every Land, Attraction, Restaurant, Shop, and Event in the Original Magic Kingdom*. Santa Monica, Calif.: Santa Monica Press, 2008.

Sullivan, Walter. "Willy Ley, Prolific Science Writer, Is Dead at 62." *New York Times*, June 25, 1969, 47.

Swanson, C. P. "Review." *Quarterly Review of Biology* 27 (March 1952): 76–77.

Thomas, Frankie Jr. "Frankie Thomas on Tom Corbett, Space Cadet." In *Earth vs. the Sci-Fi Filmmakers: 20 Interviews*, edited by Tom Weaver, 352, 357–59. Jefferson, N.C.: McFarland and Company, 2005.

Thompson, Ralph. "Books of the Times." *New York Times*, December 30, 1941, 17.

Thonssen, Lester, ed. *Representative American Speeches: 1960–1961*. New York: H. W. Wilson Company, 1961.

Thurs, Daniel P. *Science Talk: Changing Notions of Science in American Popular Culture*. New Brunswick, N.J.: Rutgers University Press, 2007.

Topham, Jonathan. "Rethinking the History of Science Popularization/Popular Science." In *Popularizing Science and Technology in the European Periphery, 1800–2000*, ed. Faidra Papanelopoulou, Agusti Nieto-Galan, and Enrique Perdiguero, 1–10. Burlington, Vt.: Ashgate, 2009.

Topsell, Edward. *History of Four-Footed Beasts and Serpents*, vols. 1 and 2. Introduction by Willy Ley. New York: Da Capo, 1967.

Uhl, Alexander H. "Lindberg Lays Foundation for Fascist Revolt." *PM*, October 5, 1941, 8.

Unknown author. "5 Youths Define Race for Space." *New York Times*, November 25, 1957, 15.

———. "Alumni Day—1948." *Technology Review*, July 1948, 506–7.

———. "Books in Brief." *Nation*, October 3, 1955, 310.

———. "Books. Space Flights." *Los Angeles Herald-Examiner*, June 9, 1968, G6.

———. "A Line O' Type or Two." *Chigago Daily Tribune*, April 6, 1949, 24.

———. "Das Raketenauto rast! Experiment oder Fahrzeug der Zukunft?" *Vorwärts*, May 23, 1928, 15.

———. "Expect Reds to Put Man in Orbit First." *Daily Defender*, February 25, 1959, 2.

———. "Expert Expects Manned Space Station by '65." *Los Angeles Times*, February 24, 1959, D19.

———. "Hi-yo, Tom Corbett!" *Newsweek*, April 2, 1951, "Radio Television."

———. "It Is X-Hour Minus 5." *New York Times*, January 15, 1950, BR7.

———. "Probekapital aus Ley: Die Fahrt ins Weltall." *Die Rakete*, April 1929, 60–61.

———. "Reading for Pleasure: How to Visit Mars." *Wall Street Journal*, June 22, 1956, 6.

———. "Rocket Auto Tops 2 Miles a Minute: In Two Seconds von Opel Reaches 62 Miles an Hour at Berlin." *New York Times*, May 24, 1928, 6.

———. "Rocket Plane Fails to Soar After Take-Off: Freezing Fuel Is Blamed as Mail Experiments Prove Futile." *Washington Post*, February 10, 1936, 3.

———. "Satellite Called Stride in Space: Experts Aware of Its Possibility." *New York Times*, July 30, 1955, 9.

———. "Science Fiction Bookshelf." *Startling Stories*, November 1951, 142.

———. "Space Exposition Is Staged in Store." *New York Times*, February 5, 1958, 11.

———. "Special Willy Ley Issue." *Monogram*, February 1959, 1–3.

———. "This Is PM." *PM*, July 4, 1941, inside cover.

———. "TWA Offering 'Flight to the Moon' Feature." *Chicago Daily Tribune*, July 10, 1955, J3.

———. "Two Rocket Plane Flights Are Hailed as First Success." *Chicago Daily Tribune*, February 24, 1936, 9.

———. "Waldemar B. Kaempffert Dies: Science Editor of the Times, 79." *New York Times*, November 28, 1956, 35.

———. "War Is at Our Doorstep—What Are We Going to o About It?" *PM Special Issue*, May 18, 1941, 5.

———. "Willy Ley Describes Your Flight to the Moon." *Sunday Midwest*, July 12, 1959, 9.

———. "Willy Ley, 62, Space Travel Expert, Dies." *Los Angeles Times*, June 25, 1969, 12.

———. *Official Guide Book of the New York World's Fair 1939*. New York: Exposition Publications, 1940.

Valier, Max. *Die Sterne Bahn und Wesen: Gemeinverständliche Einführung in die Himmelskunde*. Leipzig: R. Voigtländer, 1926.

———. *Der Vorstoss in den Weltenraum: Eine wissenshaftlich-gemeinverständliche Betractung*. München: R. Oldenbourg, 1924.

Vinen, Richard. *A History in Fragments: Europe in the Twentieth Century*. Cambridge, MA: Da Capo, 2000.

Von Engelhart, Dietrich. "Romanticism in Germany." In *Romanticism in National Context*, ed. Roy Porter and Mikulas Teich, 109–33. Cambridge: Cambridge University Press, 1988.

W., K. "Fire and Poison from the Air." *New York Times*, January 18, 1942, BR8.

Walker, Mark. *Nazi Science: Myth, Truth, and the German Atomic Bomb*. Cambridge: Perseus Publishing, 2008.

Walls, Laura Dassow. *The Passage to Cosmos: Alexander von Humboldt and the Shaping of America*. Chicago: University of Chicago Press, 2009.

Walt Disney Productions. *Man in Space* (adapted for school use by Willy Ley). Syracuse, N.Y.: L. W. Singer and Company, 1959.

———. *Man and Weather Satellites*. Syracuse, N.Y.: L. W. Singer, 1959.

———. *Mars and Beyond: A Tomorrowland Adventure*. Syracuse, N.Y.: L. W. Singer Company, 1959.

———. *Tomorrow the Moon* (adapted for school use by Willy Ley). Syracuse, N.Y.: L. W. Singer and Company, 1959.

Walters, Larry. "TV Ticker." *Chicago Daily Tribune*, July 20, 1957, C4.

Wang, Zuoyue. *In Sputnik's Shadow: The President's Science Advisory Committee and Cold War America*. New Brunswick, N.J.: Rutgers University Press, 2008.

Ward, Bob. *Dr. Space: The Life of Wernher von Braun*. Annapolis, Md.: Naval Institute Press, 2005.

Ward, Janet. *Weimar Surfaces: Urban Visual Culture in 1920s Germany*. Berkeley: University of California Press, 2001.

Weart, Spencer. *Nuclear Fear: A History of Images*. Harvard: Harvard University Press, 1988.

White, E. L. "Review." *Review of Biology* 30 (December 1955): 383.

Winter, Frank. "Frau im Mond: Fritz's Lang's Surprising, Silent Space Travel Classic." *Starlog*, January 1981, 39-41, 62.

———. *Prelude to the Space Age: The Rocket Societies, 1924–1940*. Washington, D.C.: Smithsonian Institution Press, 1983.

Wolfe, Audra J. *Competing with the Soviets: Science, Technology, and State in Cold War America*. Baltimore: Johns Hopkins University Press, 2012.

Yarber, William L. "Review." *American Biology Teacher* 31 (April 1969): 270–72.

Zell, Theodor. *Neue Tierbeobachtungen*. Stuttgart: Franckh, 1919.

———. *Moral in der Tierwelt*. Stuttgart: Franckh, 1920.

———. *Straußenpolitik: Neue Tierfabeln*. Stuttgart: Franckh, 1907.

Zirkle, Conway. *The Death of a Science in Russia*. Philadelphia: University of Pennsylvania Press, 1949.

———. "Review." *Isis* 40, no. 3 (August 1949): 290.

———. "Review." *Isis* 47, no. 4 (December 1956): 433–34.

Index

Fritzsche, Peter, 28
Frontiers, Ley's articles for, 96
"Futurama," at 1939 World's Fair, 101, 102
Future biographer, Ley's advice for, 250
"Future of the Robot Bomb" (Ley), 130
Future space explorers, dedication to, last revision of Rockets, 238
Futurism, cultural representations of space-flight and, 256

Gail, Otto Willi, 57
Galaxy magazine: Ley's articles for, 182–83, 197–98, 219, 221; Ley's contract as science editor for, 181–82
Galaxy Science Fiction, Ley's articles published in, 95
Galileo Galilei, 99, 127, 157, 228–30
Gardner, Martin, 153
Gardner, Thomas, 190
Geisel, Theodor ("Dr. Seuss"), 115
General Mills, 201, 202
"Geography for Time Travelers" (Ley), 96
German nationalism: aviation and, 28; rocket enthusiasts and, 30
German popular science, influence on Ley, 6
German Rocket Society, 88
German Student Association, 67
Germany: attack on "Jewish physics" in, 71; Bölsche and founding of "peoples' school" in, 23; death of science in, 77; militarization of rocketry in, 65–71; Nazi seizure of power in, 33, 54; persecution of scientists and engineers in, 56; post-war inflation in, 25; working students during 1920s in, 26
Gernsback, Hugo, 67, 92
Gessner, Conrad, 43–44, 107, 234
Gestapo, establishment of, 67
Gilman, William, 234
Gingerich, Owen, 230
GIRD, Ley's exchanges with, 56, 263n1
Glacial cosmogony, 68
Glenn, John, 239
Gloria (mail-carrying rocket), public experiments with, 89–90
Goddard, Robert H., 40, 87, 91, 95, 149, 154, 210; Ley compares Oberth with, 31–32; Ley on secrecy of, 63; Pendray's obituary of, 150

Goebbels, Josef, 70, 123
"Golden Age," of science fiction writers, 181
Golinski, Jan, 245
Great Depression, 59, 61
Great War. See World War I
Greek mythology, Ley's lifetime fascination with, 15
Greeks: in Dawn of Zoology, 241; in Watchers of the Skies, 226
Grissom, Gus, 239
Guinn, Robert, 181

Haber, Heinz: Collier's magazine space travel issues and, 176; Disney deal and, 185
Hall, Marie Boas, 236
Harnessing Space (Ley, ed.), 237
Harper's magazine, Pendray's articles in, 148
Hayden Planetarium (New York City), 163, 170, 175–76
Haynes, Roslyn, 17
Heinlein, Robert A., 151; excerpts from Ley's letters to, 34, 115, 116, 120, 124, 141, 142, 143, 144, 146–47, 160, 163, 169, 170, 206; generous review of Conquest of Space by, 167; Ley's visit with, 119; as technical adviser for Destination Moon, 166
Hellman, C. Doris, 231
Herbert, Don, 169
"Here He Is—the German Who Perfected Hitler's Flying Bomb" (Ley), 271n31
Herodotus, 241
Heroic Age of American Invention, The (de Camp), 233
Heroism: astronauts and, 256; engineers and, 19–20; explorers and, 23–24; science writers and, 20–23
Herschel, John, 127
Hevelius, 127
Hildegardis de Pinguia, 242
Himmler, Heinrich, 71, 140
Hindenburg, Paul von, election of, 27
Hines, William, 192, 193
Hipparchus, 227
Historia naturalis (Pliny the Elder), 241
History of science, 13, 222–23, 232; case studies in, 189; criticisms of popular writers of, 233, 234; critiqued as

History of science—*continued*
"pseudo-scholarly," 235; *Dragons in Am-ber* and, 173; *Isis* contributors' derision for popularization of, in film, 234–35; Ley on, 100; Ley's contribution to, 112, 158; in Ley's *Rockets*, 126–28; Ley's sec-ond opus on, 240; patterns in, according to Ley, 108; professionalization of, 220, 244–45; seeking broad perspective on, 246; usefulness of, for broad camp of intellectuals, 157–58

Hitler, Adolf, 33, 34, 62, 64, 129, 130, 137, 139, 140, 152, 202, 217; assumes dictato-rial powers, 67; declares war on the United States, 120

Hoaxes, 127

Hohmann, Walter, 29, 46

Hollinger, David A., 132

"Hollow Earth Doctrine," 72

Holmes, Frederic L., 233

Homer, 173

Hopalong Cassidy, 170

Hörbiger, Hans, 36, 72, 73

Horseshoe crab, Ley's interest in, 88

Houston Post, 207, 240

Howe, Quincy, 123

How the Cold War Transformed the Philoso-phy of Science (Reisch), 245–46

Hoyle, Fred, 236

Hubbard, L. Ron, 181

Hübner, Margot, failed marriage to Willy Ley, 97–98

Human agency, importance of, 255

Humboldt, Alexander von, 4, 26

Huxley, Julian, 174

I Aim for the Stars (film), 209

Individualism, astronaut and, 256

Industrialized warfare: German World War I generation of fathers and, 14; loss of chivalry due to, 18

"Infotainment," 6

Ingersoll, Ralph, 114

"Inside the Atom" (Ley), 144

Intellectual émigrés, elitist perceptions of mob irrationalism and, 111

"Interception of Long-Range Rockets, The" (Ley), 154

Intercontinental ballistic missile (ICBM), Russian announcement about, 204

International Geophysical Year, successful launch of American satellites during, 197

Internationalists, as enemies of the state, in Germany, 71

Interplanetary Bridges (Anton), 57

Intuition, triumph of reason and empiri-cism over, 3

Invisible Man, 255

Irrationalism, 72, 73, 111, 112, 131

Isis, 155, 159, 189, 231, 233, 243; contribu-tors' derision for popularizing history of science through film in, 234–35; political biases critiqued in, 235; popu-lar science writers criticized in, 233–34; purpose of, 131

Jackson, Joseph Henry, 174

Japanese Americans, World War II and internment of, 121

Jewett, Andrew, 136

Jewish physics, attack on, in Germany, 71

Joravsky, David, 235

Journey into Space (Ley), 31–33, 35, 46

Jünger, Ernst, 18, 19

Jupiter missile, 204

Kaempffert, Waldemar, 129; articles by, 133–34; obituary, 132–33, 271n35

Kant, Immanuel, 21

Kaplan, Joseph, 176, 200

Keenan, Joseph H., 156

Kent, Carleston, 192, 193

Kepler, Johannes, 35, 127, 228

Kessler, Frido W., 89, 90, 91

Kilgore, De Witt Douglas, 251–52

Killen, Andreas, 29

Killian, James R., Jr., 155, 156

Kimball, Ward, 183, 184

Knowledge, Ley and democratization of, 7

Koestler, Arthur, 234, 245

Kosmos (von Humboldt), 21

Koyré, Alexandre, 235

Krutch, Joseph Wood, 240

Kuhn, Thomas S., 231, 232, 233, 245

Ku Klux Klan, 202

Lands Beyond (Ley & de Camp), 178–80, 184

Lang, Fritz, 47, 48, 49, 50, 51, 78, 249;

Mann, W. M., 159
Man's Most Dangerous Myth: The Fallacy of Race (Montagu), 132
Mariner IV to Mars (Ley), 238
Marks, Lionel S., 149
Mars, 127, 128, 196, 207
Mars, Planet of War (Ley), 35
Mars and Beyond: A Tomorrowland Adventure (Ley), 215
Masculinity, World War I and crisis of, 18, 19
Mass media: flourishing of popular science and, 7; Ley's embrace of, 112
Mass production, 27
May, Mark A., 113
McCurdy, Howard E., 161, 183
Mead, Margaret, 132
Mechanix Illustrated, 124, 141; Ley's articles in, 129, 130, 145, 274n44
Mediterranean Sea, Ley's plan for partial draining of, 186, 187
Mein Kampf (Hitler), 33, 123
Meteorological rockets, Ley and, 141, 142, 143
Method of Reaching Extreme Altitudes, A (Goddard), 31, 63
Metropolis (film), 47
Meyer, M. Wilhelm, 20
Mickey Mouse Club, The, Sugar Jets Cereals marketing campaign and, 200
Mike Wallace Show, The, 202. *See also* Wallace, Mike
Militarism, German science fiction and, 57
Militarization of rocketry, 65–71
Miller, Ron, 160
"Mirak" rockets, testing of, 60
Missile gaps, Ley speaks about, 210–12
Missiles, American, and "shock" of *Sputnik*, 203–5
Missiles, Moonprobes, and Megaparsecs (Ley), 238
"Missiles, Moons, and Space Ships" (Ley), 207
MIT, Cold War rhetoric and broader mission of, 156–57
Modern Arms and Free Men (Bush), 157
Modernism, 13
Modernity, "Janus-faced," 3, 5
Modern technologies, Ley's confidence in, 188

Monogram space models, Ley as consultant for, 213–15
Monroe, Burt L., 190
Montagu, M. F. Ashley, 132, 155
Moon hoax, 127
Moon landing, 251; Willy Ley's death just weeks before, 247–48
Moon trips, Ley's predictions for, 209
Morse, Philip M., 155
Moskowitz, Sam, 25, 219, 230
Museum of Natural History (Berlin), Ley's thorough exploration of, 20
Mystical philosophy, 71
Mysticism, 72, 111

NASA, 219, 238, 240, 248; creation of, 205, 212; narration of recent publicity video for, 252
Nation, 163, 190
National Association of Science Writers, 134
National Council of Jewish Women, 74
National Cultural Chamber Act, 33
National Defense Education Act, 205
Nationalism, 53, 77; German science fiction and, 57; rocket fad in Weimar Germany and, 41–42
Nationalistic science, scientific internationalism in Weimar Germany and, 45
National Review, 208
National Science Foundation, 205
National Zoological Park, 159
Natural History Book Club, 171
Natural History Magazine, Ley's articles for, 92, 96, 144, 274n44
Natural History (Pliny the Elder), 107
Nature, Ley promotes romantic appreciation of, 10
Naturphilosophie, 3
Navy Department, 147
Nazi engineers, former, at work for U.S. military, 141, 144
Nazi Germany: Ley's view of, 102; rise of anti-science in, 12; rise of pseudoscience in, 71–72
Nazi party, 128; Ley's membership in, 33–34, 55; seizure of power by, 33, 54; V-2 rocket and, 139
Nebel, Rudolf, 48, 59, 60, 61, 62, 64, 65, 66, 68, 69, 70, 178

Sachs, Aaron, 4
Sagan, Carl, 2, 4, 245
Salamanders and Other Wonders (Ley), 188–91
Saler, Michael, 2, 3, 254
Santillana, Giorgio de, 229
Sarton, George, 10, 131, 159, 231, 233, 242, 245
Sartonism, 13
Satellites, 200–201; artificial, 199, 237; Ley's predictions for, 209; Vanguard, 197, 208
Satellites, Rockets, and Outer Space, Ley's articles collected in, 207
Saturday Review, 163
Schaeffer, Herbert, 150
Scholarly historians, divergence between science writers and, 245
Science: critical anti-authoritarianism and, 11; death of, in Germany, 77; dynamic power of truth and, 135; "essential interrelation" of democracy and, 135; Kaempffert on ideological manipulation of, 133; Ley's positive image of, 111; Oppenheimer on universality of, 155; Soviet, historians and reevaluation of, 235; World War I and disillusionment with, 18–19
Science, 244
Science: U.S.A. (Gilman), 234
Science and Human Values (Bronowski), 204
"Science and the Civil War" (Cohen), 155
Science fiction: on Ley's contribution to field of, 248–49; right-wing, 34; Weimar Germany, 56–57
"Science in the News" articles (Kaempffert), 134
"Science in the Totalitarian State" (Kaempffert), 133
Science writers: criticisms of, 233–34; divergence between scholarly historians and, 245; heroic, 20–23; Ley's perspectives on, 112; mass media and pubic education embraced by, 7; spiritual outlook of, 11–12; stark world view of, 11; World War II and, 131–36
Scientific American, Pendray's articles in, 95
"Scientific attitude," 111

Scientific autobiographies, historians and critical appraisal of, 233–34
Scientific internationalism, in Weimar Germany, 45
Scientific journalism, rise of, 7
Scientific management, 27
Scientific Monthly, 149, 153, 163
Scientific racism, 73
Scientific Renaissance, The (Hall), 236
Scientific Revolution, 227
Scientific spirit: Ley's relentless evangelizing for, 166; scientific intellectuals and, 111
Scientific Spirit and Democratic Faith, The (Lindeman, ed.), 134
Scientific thinking, Ley's belief in power of, 117
Scientism, 13
"Scientist as adventurer," Verne's novels and, 17
Scientists: cultural producers *vs.*, 7; German, persecution of, 56
"Search for Zero, The" (Ley), 98–100
Second World War. *See* World War II
Secularism, 3
Self-sacrifice, astronaut and, 256
Seversky, Alexander P. de, 122
Shape of Things to Come (Wells), 118
Shapiro, H. L., 144
Shells and Shooting (Ley), 122
Shell shock, German students and familial effects of, 14
Shershevsky, Alexander, 48–49
Sherwood, Taylor S., 236
Sherwood, Thomas S., 158
Shot into Infinity, The (Gail), 57
Simon and Schuster, 123
Sky, Pendray's articles in, 95
Sleepwalkers, The (Koestler), 234
Smith, David, 184
Smith & Street publishers, 95
Snow, C. P., 221, 222
Society for Progressive Transport Technology (EVFV), 69
Society for Space and Travel, 68; decline of, 56, 61, 64; disbanding of, 69; founding and ideological goals of, 37; international nature of, 45; Ley's autobiographical account of, 125, 128; Ley's vice-presidency

of, 45–46, 59; Nebel's rise within ranks of, 59; rocket fad and, 41, 42; tensions between Oberth and members in, 63

Solar energy, 187

Solomon, Stan, 89

"Some Implications of the Sputniki" (Ley), 208

Somnium (Kepler), 127

"Sounding Rockets and Geoprobes" (Ley), 221

Sources: Ley's documented investigation of, 31; secondary, Ley's lifetime distrust of, 16

"Soviets Seen Far Ahead in IRBMs" (Ley), 209

Soviet Union: Ley's view of, 102; *Sputnik I* launched by, 204

Space Age, 13, 251; American, Ley as most important publicist of, 9; beginning of, for Ley, 29; del Rey on Ley and engineering of, 249; Ley as man behind the curtain of, 165; Ley as prophet of, 171; Ley as showman of, 8; Ley on his role in, 250; neo-Humboldtian worldviews and, 255; role of popular science and, 217

Spaceflight, Ley as first historian of, 2, 8

Spaceflight magazine, tributes to Willy Ley, 248

Spaceflight technologies, Ley on military potential of, 211

Space history, power of Ley's narratives in field of, 10

Space models, Ley's design of, 213–15

Space Patrol (television show), 168, 169

Space Pilots (Ley), 201–2, 239

"Space Science Pleas Ignored" (Ley), 210

"Space Travel by 1960" (Ley), 182

"Space war," Ley's prediction of, 238

Space World magazine, Ley's articles for, 220

Speaking truth to power, freethinkers and, 16

Spiritual quest, of humanity, 254

Spohr, Carl W., 66

Sputnik, 217, 256; American missiles and "shock" of, 203–5; launch of, 195, 204; public confidence in years prior to, 166

Sputnik II, launch of, 206

Stannard, Jerry, 244

Starfield Company, The (Ley), 56–58, 63

Stark, Johannes, 71

Starrett, Vincent, 159

Startling Stories, 92

"Stations in Space" (Ley), 95

Sternig, John, 200

Stone, I. F., 115

"Story of the Fish Anguilla, The" (Ley), 274n44

"Story of the Milu, The" (Ley), 274n44

Strait of Gibraltar, Ley's plan for damming of, 187

Structure of Scientific Revolutions, The (Kuhn), 232

Sugar Jets Cereals, Ley and marketing campaign for, 200–202

Sullivan, Walter, on Willy Ley's death before moon landing, 248

Swanson, C. P., 174

Taylor, Rennie, 134

Technological innovation, Ley's naïve view of, 53

Technology, World War I and disillusionment with, 18–19

Technology Review, 139; Ley's articles for, 155, 199; Vannevar Bush and overall agenda of, 155

Television: early science fiction programs on, 167–70; Ley's appearances on, 172, 198, 206, 208, 214, 220, 252–53

Ten Lost Tribes, Ley's and de Camp's take on, 180

"Terry Bull's Terrible Weapon" (Ley), 124

Third Reich, writers declare loyalty to, 33

This Week magazine, Ley's articles in, 207

Thomas, Frankie, Jr., 165, 169, 170

Thor missile, failed launches of, 203–4

Thrilling Wonder Stories, Ley's articles published in, 95

Thurber, James, 115

Timaeus (Plato), 180

Time magazine, 209

Tom Corbett, Space Cadet (television show), 167–70, 201; Ley and publicity campaign for, 170; Ley as consultant for, 169–70, 177; Ley as "co-pilot" and public appearance at NBC studios, 165; premiere of, 167

JARED S. BUSS is adjunct professor of history at Oklahoma City Community College. He was the recipient of the 2015 Guggenheim fellowship at the Smithsonian National Air and Space Museum.